T0138445

The Mangle of Practice

The Mangle of Practice

Time, Agency, and Science

Andrew Pickering

The University of Chicago Press
Chicago and London

Andrew Pickering is professor of sociology, criticism, and interpretive theory at the University of Illinois. He is the author of *Constructing Quarks* and the editor of *Science as Practice and Culture,* both published by the University of Chicago Press.

The University of Chicago Press, Chicago 60637
The University of Chicago Press, Ltd., London

Printed in the United States of America
04 03 02 01 00 99 98 97 96 95   1 2 3 4 5

ISBN: 0-226-66802-9 (cloth)
0-226-66803-7 (paper)

Library of Congress Cataloging-in-Publication Data

Pickering, Andrew.
The mangle of practice : time, agency, and science / Andrew Pickering.
p. cm.
Includes bibiliographical references and index.
1. Science—Philosopy. 2. Science—Social aspects. I. Title.
Q175.P522 1995
501—dc20   94–44546
CIP

∞ The paper used in this publication meets the minimum requirements of the American National Standard for Information Sciences—Permanence of Paper for Printed Library Materials, ANSI Z39.48-1984.

For Thomas and Alex

Contents

Preface

One good thing about writing a book is that you get two tries at the introduction. Chapter 1 is a self-contained preview of the issues that concern me, how I propose to address them, and what I think my examples show. Here I take a more biographical route. It might help in reading the book to have an idea of how it arrived at its present shape.

I have been fascinated for a long time by knowledge. Since my schooldays, I have wondered about how knowledge relates to the world—about the problematic of realism, as philosophers call it. Having gained quite a bit of knowledge, eventually as an elementary-particle theorist, in 1976 I joined the Science Studies Unit at the University of Edinburgh, where I discovered that there were other people like me. The unit was one of the centers of a small community of people developing a sociology of scientific knowledge, and I tried as best I could to join in the absorbing but very difficult arguments that swirled around that topic. I worked through several case studies of my former discipline and finally wrote a book on it, *Constructing Quarks* (1984). By the time of that book, I had come to the conclusion that it was no use trying to think about knowledge in isolation. To understand why people believed what they did, it seemed one had to understand how specific items of knowledge fitted in with the practice of their producers and users. Indeed, *Constructing Quarks* tried to display the historical development of particle physics over a twenty-year period as instantiating a simple model of scientific practice. After the book, I was left with one nagging concern. The problem I started with had not been solved. I now felt clear enough on how knowledge related to people, but not on how it related to the world (except that it was implausible to imagine that scientific knowledge literally corresponded to the inner constitution of nature). The problematic of realism was still with me.

The important development for me came when I spent 1986–87 as a member of the Institute for Advanced Study in Princeton. Shortly after my arrival, I set out to respond to an essay review of my book by Yves Gingras and S. S. Schweber (1986). They accused me, wrongly, of Duhem-Quine abuse—of mistakenly presupposing a set of arguments associated with the names of those philosophers. Thinking about this accusation led me to an appreciation of a fact that was already emphasized in constructivist sociology of science, namely that scientific culture is not a unitary, monolithic thing (say, a single big theory, as is typically taken for granted in Duhem-Quine-type arguments); that scientific culture is, in fact, an assemblage of multiple and heterogeneous elements. And I realized that I could, on this basis, at last say something about realism. An earlier study I had made of the history of quark-search experiments could be read as an analysis of knowledge production in terms of the difficult and uncertain work of making associations between heterogeneous—in this case, material and conceptual—elements of scientific culture. An essay followed, "Living in the Material World" (1989), in which I extended my previous account of the experimental work of the physicist Giacomo Morpurgo as an exemplification of a "pragmatic realist" perspective on knowledge, a perspective that deserved to be called realist precisely because it specified the nontrivial links that scientists fashion between representations and the world.

"Material World" was important to me for several reasons. First, as just mentioned, it addressed a problematic that had been with me for years. Second, thinking about cultural multiplicity and heterogeneity helped me to see what was original in the actor-network approach to science studies then being developed by Michel Callon, Bruno Latour, and John Law. I have continued to learn from the actor-network ever since. And third, on returning from Princeton to the University of Illinois, I realized that the essay could also be read as significantly extending the analysis of practice developed in *Constructing Quarks*. In the book, I had argued that scientific practice is centrally a process of modelling, but, looking back, I had failed to appreciate sufficiently the open-endedness of modelling. One can try to extend a given scientific culture in an indefinite number of ways. What the history of Morpurgo's experiments showed was that, to put it crudely, most of those ways do not work. When Morpurgo sought to extend the material and conceptual strata of his culture, the bits did not usually fit together. "Resistances" continually arose in his work relative to the material-conceptual alignments he needed to achieve to produce facts. And, from the indefinite range of possibilities, certain specific modelling vectors were singled out in his practice precisely in that they did issue in such alignments. Practice as modelling, I thus realized, has an important real-time structure, with the contours of cultural extension being determined by the emergence in time of resistances, and by the success or failure of "accommodations" to resistance. This temporal structuring of practice as a dialectic of resistance and accommodation is, in the first instance, what I have come to call the mangle of practice. The mangle and its implications and ramifications in philosophy, social theory, and historiography are what this book is about.

Having arrived at the idea of the mangle, I began to wonder about its generality. The mangle operated, as I then understood it, at a level of detail not usually accessible to empirical study, but eventually I found and started working through the examples that follow here. From this work, it appeared that the mangle could go a long way, and thus, when I set off from Illinois to spend 1992–93 in the Department of History and Philosophy of Science in Cambridge, my ambition was to write a book about it. Then one last twist entered the story. Morpurgo's quark-search experiments had been the test case for the development of my ideas since I first saw how they could lead into the problematic of realism. For the book, I intended to extend my analysis of those experiments back into their earliest phase, and there I found myself on uncertain ground. I could not persuade myself of the story that I wanted to tell of Morpurgo's early struggles to get his apparatus to work. It was possible to talk about those struggles along lines that were becoming familiar—in terms of the making and breaking of associations between multiple and heterogeneous cultural elements—but something seemed to be missing. And that something, it appeared, was material agency. In building his apparatus, Morpurgo was trying to get the material world to do something for him, and this needed to be stated out loud. Having recognized that, my project reconstituted itself.

As discussed in the text, talk of material agency has always been suspect in the sociology of scientific knowledge, but not so in the actor-network approach. There, much is made of material agency and, further, of its symmetrical relations with human agency. It was clear, then, that if I wanted to talk about material agency I had better think what I wanted to say about human agency, too. In Cambridge, that led me in all sorts of directions: into considerations of the intentional structure of human agency, of the scale of and relationships between social actors, of the disciplined nature of scientific work, and so on. At this late stage, I gained a new appreciation of writers as diverse as Michel Foucault and Michael Lynch and a heightened (but also critical) regard for the insights of the actor-network. And the book, in its turn, developed a vital interest in human and nonhuman agency, in how they temporally intertwine, and in how knowledge engages with them. I have thus ended up with a book that I never intended to write, in which an original preoccupation with knowledge has been subsumed into a wider preoccupation with human and nonhuman agency, and in which, as explained at the end, science itself appears as subsumed within a wider field of machinic production and destruction. I am no longer puzzled by how scientific knowledge relates to the world, nor by how it relates to scientific practice; now I feel the need to understand the disciplined, industrialized, and militarized, technoscientific world in which I have lived my life, and how it ever got to be this way.

Thus the history of the mangle. I hope it will help readers come to grips with the form and content of the chapters that follow.

It remains for me to acknowledge my many debts. As mentioned above, this book began in Princeton, at the Institute for Advanced Study. It also ended in Princeton, but at the university. In 1993–94, I completed and made final revisions to the manuscript as a fellow of the Shelby Cullom Davis Center in the Princeton History Department. In between, much of the work was done at the University of Illinois at Urbana-Champaign, and for their friendship, support, and critical acumen I thank my colleagues in the Sociology Department there, my students (especially those who took part in seminars where I stumbled through the earliest versions of arguments presented here), and, above all, the members of the informal seminar of the History, Philosophy, and Sociology of Science graduate program. The HPSS seminar was the visible manifestation of a social and intellectual community without which my work would have been impossible. Also at Urbana, participation in the Unit for Criticism and Interpretive Theory was my introduction to great swathes of contemporary thought, and the unit has to take its share of responsibility for the wilder ideas expressed in what follows. Two periods of support from the History and Philosophy of Science section of the National Science Foundation marked watersheds in my project. A grant in 1989–90 helped me find the time to escape from the detailed analysis of scientific practice and to start thinking about its implications, and a second NSF grant enabled me to spend my sabbatical in Cambridge in 1992–93.

Now for the tricky part. More people than I deserve have helped me along the way to this book, reacting to talks, essays, chapters, entire manuscripts, sharing their ideas, yawning, telling me when I was talking rubbish, in correspondence, offices, seminar rooms, restaurants, homes, bars, and pubs. Increasingly over the past few years I have counted myself exceptionally fortunate that my thought and writing has been situated in a very rich and stimulating field of conversations. I therefore have no confidence whatsoever that I can generate a complete list of the individuals to whom I should express my gratitude. But still I should try, and I hope to be forgiven for what will no doubt prove to be appalling omissions. In footnotes to the text, I acknowledge specific debts, but here I offer my thanks in general to Susan Abrams, Brian Baigrie, Davis Baird, Barry Barnes, David Bloor, Geof Bowker, Nancy Cartwright, Soraya de Chadarevian, Harry Collins, Natalie Davis, Norman Denzin, Irving Elichirigoity, Paul Feyerabend, Owen Flanagan, Paul Forman, Peter Galison, Dilip Gaonkar, Gerry Geison, Yves Gingras, Laurel Graham, Ian Hacking, Mary Hesse, Joann Hoy, Piet Hut, Robert Alun Jones, Vera Ketelboeter, Yiannis Koutalos, Martin Krieger, Thomas Kuhn, Bruce Lambert, Michèle Lamont, Bruno Latour, John Law, Peter Lipton, Michael Lynch, Michael Mahoney, Peter Miller, Giacomo Morpurgo, Malcolm Nicolson, Ted O'Leary, Ronald Overmann, Trevor Pinch, Michael Power, Diederick Raven, Joseph Rouse, Simon Schaffer, Sam Schweber, Steven Shapin, Otto Sibum, Barbara Herrnstein Smith, Betty Smocovitis, Leigh Star, Adam Stephanides, Fred Suppe, Peter Trower, Stephen Turner, Adrian Wilson, Norton Wise, and Alison Wylie.

From that list, I should single out two people. Since our days together at the Institute for Advanced Study, Barbara Herrnstein Smith has offered me consistently perceptive advice and incisive criticism. The trajectories of our research run along intersecting lines; I have learned much from her writings, and even more in conversations and arguments with her. The other person is Simon Schaffer. At the beginning, it was he who encouraged me to write "Living in the Material World." Almost at the end, Simon, as acting head of department, made possible my sabbatical in Cambridge and, together with Anita, helped make it such a pleasure. While I was there, he somehow found time to read and comment upon not one but two quite different drafts of this book. And more than that, he typically displayed a clearer grasp of where I was going and how to get there than I did. It was an education to have the chance of interacting with him over an extended period of time. So, thanks, Simon and Barbara. Thinking of Cambridge, I also want to express my gratitude to my old friends Jim and Rhonda and Lee and Paula for their help above and beyond the claims of friendship in the hassles of transplanting me, Jane, and the children back to our native land. I recall remarking on several occasions, and only partly in jest, that they had saved our lives. Most of my writing in 1992–93 was done at Lee and Paula's old dining table, though Paul Waldmann's carpenter's bench deserves a mention, too.

Lastly, as ever I thank Jane F.—oh, for everything.

I think by writing, and my route to this book has been marked by a trail of essays, first stabs at stories and analyses that, for the reasons described above, especially my newfound concern with questions of agency, appear here and there in the text, transformed, redistributed, and accompanied by new material. The essays that have seen print, and the chapters to which they relate, are "Living in the Material World: On Realism and Experimental Practice," in D. Gooding, T. J. Pinch, and S. Schaffer, eds., *The Uses of Experiment: Studies of Experimentation in the Natural Sciences* (Cambridge: Cambridge University Press, 1989), pp. 275–97 (chap. 3); "Knowledge, Practice, and Mere Construction," *Social Studies of Science* 20 (1990): 682–729 (chaps. 6, 7); "Objectivity and the Mangle of Practice," *Annals of Scholarship* 8 (1991): 409–25 (chap. 6); "Constructing Quaternions: On the Analysis of Conceptual Practice," co-authored with Adam Stephanides, in Pickering, ed., *Science as Practice and Culture* (Chicago: University of Chicago Press, 1992), pp. 139–67 (chap. 4); "Anti-Discipline or Narratives of Illusion," in E. Messer-Davidow, D. Shumway, and D. Sylvan, eds., *Knowledges: Historical and Critical Studies in Disciplinarity* (Charlottesville: University Press of Virginia, 1993), pp. 103–22 (chap. 7); "The Mangle of Practice: Agency and Emergence in the Sociology of Science," *American Journal of Sociology* 99 (1993): 559–89 (chaps. 1, 2); and "Beyond Constraint: The Temporality of Practice and the Historicity of Knowledge," in J. Buchwald, ed., *Scientific Practice: Theories and Stories of Physics* (Chicago: University of Chicago Press, 1995) (chaps. 2, 5, 6).

ONE

The Mangle of Practice

[T]here is at all times enough past for all the different futures in sight, and more besides, to find their reasons in it, and whichever future comes will slide out of that past as easily as the train slides by the switch.

William James, *The Meaning of Truth*

Desire only exists when assembled or machined. You cannot grasp or conceive of a desire outside a determinate assemblage, on a plane which is not pre-existent but which must itself be constructed . . . In retrospect every assemblage expresses and creates a desire by constructing the plane which makes it possible and, by making it possible, brings it about . . . *[Desire] is constructivist, not at all spontaneist.*

Gilles Deleuze and Christine Parnet, *Dialogues*

This is a book about science that ventures into the worlds of mathematics, technology, and the workplace. It offers a general analysis of scientific practice, which I call the mangle, and some pointers as to how it might be extended toward an understanding of the reciprocal production of science, technology, and society (STS).[1] It is also a book about time and agency that addresses central questions in the philosophy, social theory, and historiography of science and beyond. This chapter lays out some basic features of my position; the rest of the book consists of examples and articulations.

1.1 SCIENCE AS PRACTICE AND CULTURE

Science studies has been an exciting field over the past few decades, and one source of this excitement has been a continual expansion of conceptions of science as an object of study.[2] Until the late 1950s, it seemed

1. I hope, therefore, that "science" will be read hereafter as an umbrella term of a greater than usual extent.

2. The core fields of "science studies" are history, philosophy, and sociology of science but, as indicated in the previous note, I construe the term broadly to encompass both

enough to think of science as a body of knowledge, a collection of empirical and theoretical propositions about the world. This body of knowledge constituted a self-contained topic for the philosophy of science, for example, whose job it was to enquire into, and possibly to legislate upon, the formal relations between the propositions it contained. But the work of Norwood Russell Hanson (1958), Thomas Kuhn (1970) and Paul Feyerabend (1975) changed all that.[3] Especially Kuhn's persuasive periodization of the history of science into stretches of "normal science" separated by "revolutionary" gulfs challenged the self-containment of the science object, and opened the way for new waves of scholarship to wash over and reconceive it. Thus, since the 1970s, work on the sociology of scientific knowledge (SSK) has increasingly documented the importance of the human and the social in the production and use of scientific knowledge.[4] Social structure, social interests, human skills—all of these have come to be seen as constitutive of science, as integral to science in interesting and important ways. Further, though SSK's primary focus has been, as its name states, on knowledge and the social, empirical work in SSK has served also to foreground the material dimension of modern science—the omnipresence of machines, instruments, and experimental setups in scientific research. This dimension had long been ignored in mainstream history and philosophy of science, but here SSK has made contact with an alternative philosophical perspective powerfully articulated by Ian Hacking (1983), which seeks precisely to emphasize the machinic aspects of science. Finally, from the late 1970s to the present, scholars have evinced an increasing interest in the details of the day-to-day doing of science. This interest has served further to expand our conception of the science object by documenting its sheer multiplicity and heterogeneity. All of the dimensions of science

---

contributions from other disciplines interested in science (anthropology, political science) and work that extends into the study of mathematics and STS.

3. Suppe (1977) surveys developments in philosophy of science consequent upon the work of these authors.

4. Canonical books in SSK include Barnes 1974, 1977, 1982; Bloor 1991, 1983; Collins 1992; and MacKenzie 1981b. See also Shapin (1982) for an excellent review of the history and sociology of science literature from an SSK perspective. I should explain why I single out SSK here, rather than the preexisting Mertonian approach to the sociology of science (for a comparative review of both, see Zuckerman 1988). Although the latter did expand our conception of science as an object of study by exploring its institutional structure, it did not envisage the detailed intertwining of the social with the technical in science that is a major concern in both SSK and this book. The Mertonian approach, as it is said, has been more a sociology of scientists than of science.

just mentioned—the conceptual, the social, the material—have to be seen as fragmented, disunified, scrappy.[5]

I can now sketch out the problematic of this book by, first, reexpressing what has just been said in terms of an expansion of our concept of scientific *culture*. Whereas one could once get away with thinking of scientific culture as simply a field of knowledge, in what follows I take "culture" in a broad sense, to denote the "made things" of science, in which I include skills and social relations, machines and instruments, as well as scientific facts and theories. And then I can state that my abiding concern is with scientific *practice,* understood as *the work of cultural extension*. My problematic thus includes the traditional one of understanding how new knowledge is produced in science, but goes beyond it in its interest in the transformation of the material and social dimensions of science, too.

Two points of clarification should immediately be entered. One is that in this book I seek a *real-time* understanding of practice. I want to understand the work of cultural extension in science as it happens in time. This is to be contrasted with retrospective approaches that look backward from some terminus of cultural extension and explain practice in terms of the substance of that terminus. The exemplary instance of the latter is what I call "the scientist's account" (Pickering 1984b), in which accepted scientific knowledge functions as an interpretive yardstick in reconstructing the history of its own production. I think that there are serious historiographic problems in such retrospective accounting for science (Pickering 1989b), but rather than rehearse them here, let me just note that my project is a different one and that, for my purposes, to indulge in retrospection would be circularly self-defeating and must be eschewed.

My second point of clarification takes us back to the recent history of science studies. It is probably true to say that many authors engaged in exploring the work of science once shared my interest in time—back

5. Several streams of work helped to constitute the doing of science as a topic for research in its own right. Within traditional history of science, Holmes (1974, 1981, 1985) has pursued the theme most tenaciously (an important recent study is Kohler 1994), but perhaps the most influential source has been the development of ethnographic studies of "laboratory life": Latour and Woolgar 1986; Knorr-Cetina 1981; Lynch 1985a. From another angle, the interest in scientific work appears as a continuation of the concerns of SSK: Pickering 1984b; Gooding 1990. Yet another source has been pragmatist studies of science (which take the work of Howard Becker and, especially, Anselm Strauss as their point of departure): see Star 1991b, 1992; and Fujimura 1992. For surveys of current perspectives, see Pickering 1992b; and Clarke and Fujimura 1992.

in the early 1980s, say. But as I have been writing this book, it has dawned on me that a kind of purification has taken place. Much of the most interesting work now being done is not concerned with practice as I have just defined it, but takes the form of atemporal cultural mappings and theoretical reflections thereon. My present interest in the temporality of cultural extension leaves me, I think, in a minority as far as current initiatives in science studies are concerned.[6] I say this not in a spirit of critique of what I call the cultural-studies approach—in fact, I draw extensively upon its findings as the book goes on, and I argue in section 7.2 that it is complementary to my own approach—but to make clear the tendency of this book. And in this connection, I think it will help at this stage to note an ambiguity in the word "practice," an ambiguity that explains why it took me, at least, so long to realize that my project had diverged from others.

One sense of "practice" is the generic one around which all that follows is organized—practice as the work of cultural extension and transformation in time. The other sense of "practice" relates to specific, repeatable sequences of activities on which scientists rely in their daily work—things like the "plasmid prep" in molecular biology, discussed by Jordan and Lynch (1992). Unlike my generic sense of "practice," this one has a plural form; one can talk, for instance, of a distinct set of practices as characteristic of a given science or laboratory. And—this is the important point—*practices,* on my definition, *fall into the sphere of culture,* and the study of practices in their own right falls into the domain of what I just called cultural studies. In contrast, I am interested in practices not so much in themselves but inasmuch as they are among the resources for scientific practice and are transformed (or transformable) in practice, alongside all of the other components of scientific culture. I return to the practice/practices distinction in section 1.3, but for now the simple message is that thinking about science bifurcates in the

6. I cannot offer statistical evidence on this point, and it would not affect my arguments in the rest of the book if I were wrong. But still, my reading of the latest works of the science-studies authors I most admire tells me that they are largely engaged in projects different from my own. As an example of this, when I look back on the book I recently edited, *Science as Practice and Culture* (1992), I see now, as I did not then, that only Gooding's essay there shares my consistent interest in cultural transformation in time. The actor-network approach to science studies discussed in section 1.3 has been very important to my own thinking about the temporality of practice, but has largely been developed and appropriated by others in a cultural-studies mode, perhaps because many of its key terms—"centers of calculation," "obligatory passage points" as well as "network" itself—are terms of art in cultural mapping.

act of deciding whether "practice" has a plural or not. In this book, I take the latter fork.

1.2 REPRESENTATION AND PERFORMATIVITY

Satisfied that the sequence of men led to nothing and that the sequence of their society could lead no further, while the mere sequence of time was artificial, and the sequence of thought was chaos, he turned at last to the sequence of force; and thus it happened that, after ten years' pursuit, he found himself lying in the Gallery of Machines at the Great Exposition of 1900, his historical neck broken by the sudden irruption of forces totally new.

Henry Adams, *The Education of Henry Adams*

Before I get into the details of my understanding of scientific practice, I need to talk about the metaphysics that informs it. In particular, I want to contrast what I call the representational and performative idioms for thinking about science. The representational idiom casts science as, above all, an activity that seeks to represent nature, to produce knowledge that maps, mirrors, or corresponds to how the world really is. In so doing, it precipitates a characteristic set of fears about the adequacy of scientific representation that constitute the familiar philosophical problematics of realism and objectivity (which I discuss in chapter 6). Of course, within the traditionally restricted vision of science-as-knowledge, the representational idiom is more or less obligatory—what else can one ask of knowledge other than whether it corresponds to its object?–but it has continued strongly to inflect our understandings of science even to the present. It has been taken to a new pitch of intensity, for example, in the reflexive approach to science studies developed in the works of Michael Mulkay (1985), Steve Woolgar (1988b, 1988c), and Malcolm Ashmore (1989). Reflexivity turns the usual fears concerning the adequacy of representation—the "methodological horrors," as Woolgar calls them—back upon science studies itself, but never doubts that the point of science and science studies is indeed representation.[7]

Within an expanded conception of scientific culture, however—one

7. Woolgar (1988a, 1992) makes clear the connection between the reflexive turn in science and that taking place more generally in the human sciences, especially in anthropology. The grip of the representational idiom is exemplified in Lynch's essay "Representation Is Overrated" (1993a). Lynch's conclusion is that we need to study representation differently, not that we should escape from the representational idiom.

that goes beyond science-as-knowledge, to include the material, social, and temporal dimensions of science—it becomes possible to imagine that science is not just about representation. And working through the studies discussed later in this book has convinced me that it is both possible and necessary to escape from the representational idiom if we are to get to grips with scientific practice. The point is this: Within the representational idiom, people and things tend to appear as shadows of themselves. Scientists figure as disembodied intellects making knowledge in a field of facts and observations (and language, as the reflexivists remind us). But there is quite another way of thinking about science. One can start from the idea that the world is filled not, in the first instance, with facts and observations, but with *agency.* The world, I want to say, is continually *doing things,* things that bear upon us not as observation statements upon disembodied intellects but as forces upon material beings. Think of the weather. Winds, storms, droughts, floods, heat and cold—all of these engage with our bodies as well as our minds, often in life-threatening ways. The parts of the world that I know best are ones where one could not survive for any length of time without responding in a very direct way to such material agency—even in an English summer (never mind a midwestern winter) one would die quite quickly of exposure to the elements in the absence of clothing, buildings, heating, and whatever. Much of everyday life, I would say, has this character of coping with material agency, agency that comes at us from outside the human realm and that cannot be reduced to anything within that realm.[8]

My suggestion is that we should see science (and, of course, technology) as a continuation and extension of this business of coping with

8. I thank Malcolm Nicolson and Soraya de Chadarevian for emphasizing to me the relation between these remarks and the phenomenological tradition in Continental philosophy. I confess to little knowledge of the works of the masters, Husserl, Heidegger, and Merleau-Ponty, though Rouse (1985, 1986, 1987a) makes fascinating connections in seeking to inject these authors into contemporary Anglo-American philosophy of science. See also Rouse 1987b; Nicolson 1991; and note 27 below. I also thank Simon Schaffer and Steven Shapin for pressing me on my notion of material agency. They alerted me to the idea that concepts of agency and intentionality are bound together, and that therefore agency can properly be attributed only to human beings. I seek to clarify this issue below; in the meantime, I note that I am not alone in finding it useful and appropriate to speak of material agency. Thus, for example, Harré and Madden (1975) are happy enough to speak of agency in nature without ever dreaming of imputing intentionality to it, and Wise and Smith (1989–90, 419) quote William Whewell, writing in 1841 that "[i]n many cases the work to be done may be performed by various agencies; by men, by horses, by water, by wind, by steam."

material agency. And, further, we should see *machines* as central to how scientists do this. Scientists, as human agents, maneuver in a field of material agency, constructing machines that, as I shall say, variously capture, seduce, download, recruit, enroll, or materialize that agency, taming and domesticating it, putting it at our service, often in the accomplishment of tasks that are simply beyond the capacities of naked human minds and bodies, individually or collectively.[9] A windmill grinds grain very much faster than a miller could do by hand; my television set shows me events distant in time and space that I could not otherwise hope to view; a machine tool cuts metal at a speed and with a precision that no one could otherwise hope to achieve.

These remarks, then, sketch out a basis for a *performative* image of science, in which science is regarded a field of powers, capacities, and performances, situated in machinic captures of material agency. And my aim in the rest of this book is to understand scientific practice within such a performative idiom. I can immediately add that thinking about material performativity does not imply that we have to forget about the representational aspects of science. Science is not just about making machines, and one cannot claim to have an analysis of science without offering an account of its conceptual and representational dimensions. Much of this book (chapters 3, 4, and 6) is therefore directly concerned with the production of scientific knowledge. The move to the performative idiom, however, does imply a certain strategy in thinking about scientific knowledge. It suggests that we should explore the many different and interesting ways in which knowledge is threaded through the machinic field of science. The performative idiom can thus include the concerns of the representational idiom; it is a *rebalancing* of our understanding of science away from a pure obsession with knowledge and toward a recognition of science's material powers.[10] The machine, as I conceive it, is the balance point, liminal between the human and nonhuman worlds (and liminal, too, between the worlds of science, technology and society). I can also add that the move from a representational to a performative idiom in thinking about science has considerable historical plausibility. While the representational idiom might seem appropriate to

9. This formulation should make clear an alignment of my approach with that of the actor-network theorists, to whom I return below. The metaphor of domestication I take in particular from Callon (1986) and Latour (1987).

10. For an analysis that emphasizes the performative role in science of language itself, see Rouse 1994; more generally, see also Smith 1993a.

classical naked-eye astronomy, most of the science of our day has evolved hand in hand with the machines that were at the heart of the industrial revolution of the eighteenth and nineteenth centuries (I return to this observation in chapter 7). Further, Martin Krieger's amazing philosophical anthropology of "doing physics" (1992) suggests that the performative idiom has a phenomenological warrant, too. Krieger's suggestion is that physicists take hold of the world as if it were a factory, a site of productive equipment that needs to be managed.

Outside the world of science studies, of course, the move from a representational to a performative idiom for thinking about science might seem obvious. As Krieger says, scientists themselves do, by and large, think about the world as a field of agency, and the machinic, material dimension of their dealings with it is hardly a secret to them. The scientific laity, too, I suspect, is less impressed by representations of nature than by the display of the machinic agency of science. The science pages of the popular press hardly distinguish between scientific theory and the latest high-tech gadgets, and devote more space to the latter. One might therefore have imagined that the representational idiom would be easily dislodged within science studies. Actually, though, there are many issues that need to be thought through in the move to a performative idiom, some of which we can now review.

First an issue that is not difficult. The weight of history is on the side of the representational idiom. Going back at least to the seventeenth century, philosophers have by and large sought to keep material agency out of their discussions of science.[11] Even within philosophy of science, though, there is an alternative approach—exemplified in the work of Ian Hacking (1983), Nancy Cartwright (1983, 1989), Davis Baird and Alfred Nordmann (1994), and others—that points to the conclusion that a restriction to the representational idiom is by no means neces-

11. Ian Hacking (1983, 46) traces the philosophical aversion to speech about agency and causes in nature back to Newton's work on gravitation: Newton provided an explanation for the motion of the planets that did not, in the terms of his age, include an account of the detailed causes of that motion. Both Bloor (1983, 152–55) and Shapin (1982, 181–84) review the literature on arguments for the passivity of matter put forward in seventeenth-century England, relating them to their social, moral, religious, and political context (without remarking that SSK itself reproduces the same image of matter; see below). Philosophical arguments for the performative rather than the representational idiom often identify David Hume as a key figure in the effacement of material agency from philosophical discourse: see Cartwright 1989, 2, for example (and, from a different angle, Husserl 1970, 84–90).

sary.[12] As just noted, the world of science happens to be quite evidently and amply stocked with material agents. Hacking's well-known slogan (1983, 23)—"if you can spray them then they are real"—refers, for example, to the use of radioactive sources to change the electric charges of samples in quark-search experiments.[13] I am not persuaded of Hacking's realism, but such radioactive sources are certainly instances of what I have in mind as material agency—they are objects that do things in the world. So the sheer weight of history and custom behind the representational idiom in philosophy need not deter us from moving toward a performative one. But there are more difficult problems that should give us pause, problems that are better approached from a sociological rather than a philosophical perspective.[14]

1.3 AGENCY AND EMERGENCE

The great achievement of SSK was to bring the human and social dimensions of science to the fore. SSK, one can say, thematized the role of *human agency* in science. It thus partially displaced the representational idiom by seeing the production, evaluation, and use of scientific knowledge as structured by the interests of and constraints upon real human agents. Scientific beliefs, according to SSK, are to be sociologically accounted for in just those terms. Thus, in his classic study of the phrenology debates in early-nineteenth-century Edinburgh, Steven Shapin (1979) argues that the differing accounts of the brain produced by the competing parties have to be understood not in terms of their correspondence, or lack of it, to how the brain really is, but in relation to the divergent social interests of the phrenologists and their establishment critics: the phrenologists wanted to reform society in ways that the establishment resisted. On this view, then, scientific representations of na-

12. See also Bhaskar 1975; Chalmers 1992; Harré and Madden 1975; and Mellor 1974. I thank Peter Lipton for introducing me to this line of philosophical thought.

13. Hacking formulates this slogan while discussing the work of William Fairbank that figures in section 6.6 below.

14. In philosophy, the best-known general critique of representationalism is Rorty 1979. Hacking 1983 is the opening salvo in philosophy of science, but the most sustained encounter with, and critique of, representationalism is Rouse forthcoming. Rorty's favored alternative is the move to a conversational idiom. The shift I recommend is more like Hacking's move from representing to intervening. Rouse recommends a cultural-studies approach to science.

ture have to be understood in terms of particular configurations of human agency.

Now, there is no denying the fruitfulness of the eruption of SSK in science studies. It has irreversibly transformed the field for the better. But there is also no denying that it stands in the way of developing a fully performative understanding of science. The problem is that SSK makes it impossible to take material agency seriously. The other side of SSK's focus on human agency is precisely the invisibility of material agency. SSK, that is, is happy to talk of *accounts* of material agency as components of scientific knowledge, but it insists that such accounts be analyzed, like any other component, by referring them back to some field of human agency. The question for SSK is always, why does some community of scientists have this account of material agency? and any appeal to material agency and performativity as part of the answer is ruled out from the start.[15] SSK's account of science is, then, at most semiperformative.

Now, why should SSK adopt this asymmetric stance concerning material and human agency? Part of the answer, I think, is that it has unreflectively taken over the representational idiom from traditional philosophy of science.[16] But part of the answer is more interesting. Within SSK, one can generate arguments that point up real difficulties in moving to a fully performative idiom that acknowledges both material and human agency. And I want now to review a debate that crystallizes those difficulties, the so-called chicken debate, launched by Harry Collins and Steven Yearley's essay (1992a), "Epistemological Chicken." Collins and Yearley's target is the actor-network approach developed in the works

15. I confess that I argued along these lines in Pickering 1984c, chap. 1; 1984a. For an extension of the standard SSK perspective into technology studies, see Pinch and Bijker 1984.

16. Thus, perhaps, Bloor (1991, 158) *regrets* the absence of material agency in his canonical articulation of the SSK position: "The shortcomings of the views developed here are, no doubt, legion. The one I feel most keenly is that, whilst I have stressed the materialist character of the sociological approach, still the materialism tends to be passive rather than active. It cannot, I hope, be said to be totally undialectical, but without doubt it represents knowledge as theory rather than practice." Interestingly, Bloor goes on to compare his own position with the philosophy of Karl Popper, who "makes science a matter of pure theory rather than reliable technique." (I thank David Bloor for drawing my attention to these quotations.) In contrast, Barnes (1991, 331) is forthright in his rejection of a role for material agency in sociological accounting: "Reality will tolerate alternative descriptions without protest. We may say what we will of it, and it will not disagree. Sociologists of knowledge rightly reject realist epistemologies that *empower* reality."

of Michel Callon, Bruno Latour, and John Law.[17] Since the early 1980s, the actor-network has, I believe, shown the way toward a fully performative understanding of science, and thus a discussion of Callon and Latour's response (1992) to Collins and Yearley can serve two purposes here. On the one hand, it offers me the occasion to emphasize that I have learned an enormous amount from the actor-network. Its traces will be evident throughout this book, in the text and footnotes, and I would happy if the book were read as an attempt at a constructive dialogue with the actor-network. On the other hand, a discussion of the points at which my analysis diverges from that of the actor-network will enable me to set out my own thoughts on time, agency, and practice.[18]

The basic metaphysics of the actor-network is that we should think of science (and technology and society) as a field of human and nonhuman (material) agency. Human and nonhuman agents are associated with one another in networks, and evolve together within those networks. The actor-network picture is thus *symmetrical* with respect to human and nonhuman agency. Neither is reduced to the other; each is constitutive of science; therefore we need to think about both at once. This "extended symmetry" of humans and nonhumans is the target of Collins

17. For recent important presentations and developments of actor-network theory, see Callon 1994; Latour 1987, 1988a; and Law 1993. Rouse 1987b is an excellent philosophical discussion and extension of the actor-network approach. I should note that Collins and Yearley's essay actually has two targets: the second is Steve Woolgar's reflexivity program. As indicated earlier, I understand reflexivity as an intensification of the representational idiom in science studies, and since I want to escape from that idiom, I will not discuss Collins and Yearley's critique of it here. One point is, however, worth clarifying. Reflexivity is, like the actor-network, symmetric about human and material agency, but in a negative rather than a positive way. Reflexivity adds to SSK's deconstruction of scientists' accounts of material agency a deconstruction of SSK's accounts of the human agency of scientists. I also note below that, when the actor-network makes its semiotic move, it appears to circle back toward reflexivity. Woolgar 1992 is his response to Collins and Yearley's critique; Collins and Yearley 1992b is their response to Callon and Latour's response (1992) to Collins and Yearley 1992a.

18. One shortcoming of this expository strategy is that it ties my analysis of practice more closely to material agency than is necessary; as explained in section 1.5, my overall analysis of practice is applicable even to situations where material agency is absent. A more specific point of divergence from the actor-network than those that follow can also be noted here, concerning the "network" metaphor itself. While it performs useful work in the classic actor-network studies, I have not found it illuminating in connection with my own studies, and it plays no role in the following chapters. Mol and Law (1994) argue for a more complex "social topology" of "regions," "networks," and "fluids," without suggesting that this is an exhaustive list. I deploy some further topological concepts in section 7.4.

and Yearley's critique, which hinges upon the following dilemma. As analysts, Collins and Yearley suggest, we have just two alternatives. We can see scientists as producing accounts of material agency, in which case these accounts fall into the domain of scientific knowledge and should be analyzed sociologically as the products of human agents. This is the standard SSK position that Collins and Yearley want to defend.[19] Or we can try to take material agency seriously, on its own terms—but then we yield up our analytic authority to the scientists themselves. Scientists, not sociologists, have the instruments and conceptual apparatus required to tell us what material agency really is. The upshot of this dilemma therefore seems to be that any sociologist with a shred of self-respect had better stick to an SSK-style analysis of scientific accounting for material agency, and had better not incorporate material agency per se into her interpretive schemes. The alternative seems to be a simple rehearsal of scientific/engineering accounts of how the world is, from which the human and the social simply disappear. One can have human or nonhuman agency, but not both. This is the basic problem in moving to a fully performative idiom, as seen from the perspective of SSK.

On behalf of actor-network theory, Callon and Latour (1992) reply to Collins and Yearley that they do not see things quite this way, and neither do I, though here our positions diverge as discussed below. Callon and Latour reject the prongs of Collins and Yearley's dilemma. They insist, rightly I think, that there are not just two alternatives in the treatment of nonhuman agency. And their position, if I have understood it correctly, is this. We should not see nonhuman agency in the terms offered us by scientists or by SSK. Instead we should think *semiotically.* Semiotics, the science of signs, teaches us how to think symmetrically about human and nonhuman agents. In texts, agents (actors, actants) are continually coming into being, fading away, moving around, changing places with one another, and so on. Importantly, their status can easily make the transit between being real entities and social constructs, and back again. Semiotics thus offers us a way of avoiding the horns of

19. A certain incongruity will become apparent as this chapter and the book goes on. I find much of Collins's writing, both before and after the chicken debate, and even the later passages of Collins and Yearley's second contribution to it (1992b), to be more consistent with the actor-network approach and the analysis of practice that I want to develop than with the pure-line SSK of Collins and Yearley 1992a. I have no compelling analysis of this inconsistency; perhaps Collins's humanism (see below) is in tension with his technical analyses. To put the point more positively, I have learned very much from Collins in developing my own account of practice.

Collins and Yearley's dilemma: the agencies we speak about are semiotic ones, not confined to the rigid categories that SSK and engineering impose.

This is a clever and ingenious response, but it brings with it two different kinds of problem. One concerns human agency, and I return to it below. The other concerns material agency. From my point of view, the most attractive feature of the actor-network approach is precisely that its acknowledgment of material agency can help us to escape from the spell of representation. It points a way to a thoroughgoing shift into the performative idiom. And from this perspective, the appeal to semiotics in the face of Collins and Yearley's dilemma looks like a kind of retreat, a return to the world of texts and representations that one does not wish to make.[20] Fortunately, as I shall now explain briefly in anticipation of the empirical examples to follow, there is another way of steering our way around the dilemma.

20. Semiotics cannot be the whole story about the actor-network understanding of nonhuman agency, but it has been a central theme since Latour and Woolgar's (1986) emphasis on scientific instruments as "inscription devices," and seems invariably to be invoked under pressure. Thus, in response to Collins and Yearley's argument, Callon and Latour draw a diagram (1992, 349, fig. 12.3) in which Collins and Yearley's dilemma is represented as a horizontal nature/society axis, to which Callon and Latour add a vertical axis on which they comment (350): "The vertical axis, however, is centered on the very activity of shifting out agencies—which is, by the way, the semiotic definition of an actant devoid of its logo- and anthropocentric connotations." Likewise, in response to Schaffer's argument (1991) that Latour (1988b) depends on an illegitimate hylozoism—that is, an unjustified imputation of agency to the nonhuman realm—Latour (1992a) offered a reading of a single memoir by Pasteur. It seems clear that Latour's route to nonhuman agency in this instance is via texts in the most literal sense. (I have only an incomplete draft of Latour 1992a, but I attended his seminar presentation of the paper—Cambridge, October 1992—and subsequent conversations with him confirmed this judgment. The substance of Latour's talk has since been published (1992b), not as a response to Schaffer but as an exemplification of the semiotic method.) In a similar vein, Callon writes: "Sociology is simply an extension of the science of inscriptions. Now it should broaden its scope to include not only actors but the intermediaries through which they speak . . . *the social can be read in the inscriptions that mark the intermediaries*" (1991, 140). The position that I take below is closer to that developed earlier by John Law in, for example, Law 1987, where, without any detours through semiotics, he invokes natural forces as part of an actor-network account of the Portuguese maritime expansion. One can also note that even when Latour claims to be doing semiotics he often seems to forget, and to speak directly about laboratory practices (rather than textual accounts thereof): see, for example, key passages in Latour 1992b. More formally, Latour's recent writings introduce a peculiar quasi-semiotic operator called "shifting down," which moves between texts and the material world—between, to mention his own example, textual representations of air travel and the real thing (1993a, 16). I think that the semiotic exposition of the actor-network starts to unravel at this point.

My alternative to the invocation of an atemporal science of signs is to think carefully about time. We can take material agency in science just as seriously as SSK takes human agency, and still avoid Collins and Yearley's dilemma, if we note that the former is *temporally emergent* in practice. The contours of material agency are never decisively known in advance, scientists continually have to explore them in their work, problems always arise and have to be solved in the development of, say, new machines.[21] And such solutions—if they are found at all—take the form, at minimum, of a kind of delicate material positioning or tuning, where I use "tuning" in the sense of tuning a radio set or car engine, with the caveat that the character of the "signal" is not known in advance in scientific research.[22] Thus, if we agree that, as already stipulated, we are interested in achieving a *real-time* understanding of scientific practice, then it is clear that the scientist is in no better a position than the sociologist when it comes to material agency. No one knows in advance the shape of future machines and what they will do, but we can track the

21. One can make a connection to early pragmatist philosophy here. Discussing the work of Charles Sanders Peirce, Cohen (1923, xix n. 11) explains that "Peirce's tychism is indebted to [Chauncey] Wright's doctrine of accidents and 'cosmic weather,' a doctrine which maintained against LaPlace that a mind knowing nature from moment to moment is bound to encounter genuine novelty in phenomena, which no amount of knowledge would enable us to foresee." The same doctrine is expressed in William James's well-known sentiment that "[e]xperience, as we know, has ways of *boiling over,* and making us correct our present formulas" ([1907, 1909] 1978, 106).

22. The notion of tuning is close to Knorr-Cetina's notion (1981) of tinkering, except that the former immediately invokes the otherness of material (and other kinds of) agency. As noted in chapter 3, it was in thinking about the tuning of material apparatus that I became convinced of the need to include material agency in my analysis of scientific practice. Fleck ([1935] 1979) discusses the tuning (in my terms) of the Wassermann reaction as a test for syphilis. He notes that "[d]uring the initial experiments it produced barely 15–20 percent positive results in cases of confirmed syphilis" (72), but that after a period of collective development of the detailed performance of the reaction the success rate rose to 70–90 percent. Collins's account (1992, chap. 3) of the work of building a TEA laser can likewise be read as an ethnographic exemplification of the tuning of a material instrument. For more on the tuning of experimental devices in science, see Pickering (1984b, 14 and passim) and the works cited there. Note that it is crucial that we are considering scientific practice as the work of extending, rather than reproducing, scientific culture—in the sense of building new machines and so on. I argue that material agency is temporally emergent in relation to practice so conceived. Whether material agency per se is temporally emergent is another matter. The relative reliability of certain machines—the fact that some magnets, cars, and TVs perform the same functions day after day, for example—indicates that some aspects of material agency evolve at most slowly on the time scale of human affairs. I return to this topic in section 7.6; and I thank Adrian Wilson for prompting me to think about it in the first place.

process of establishing that shape without returning to the SSK position that only human agency is involved in it. Of course, after the fact, scientists often offer persuasive technical accounts of why the machinic field of science has developed in specific ways. But, as I said earlier, for the purposes of real-time accounting, the substance of such retrospective accounts is one aspect of what needs to be analyzed; it would make no sense to bow to the scientists and incorporate their retrospection as part of our explanation. This is my basic thought on how to think about the role of nonhuman agency in scientific practice.

Now for the second difficulty with the actor-network's semiotic move, the one that centers on human agency. Semiotics imposes an exact symmetry amounting to an equivalence and *interchangeability* between the human and material realms. Semiotically, as the actor-network insists, there is no difference between human and nonhuman agents. Semiotically, human and nonhuman agency can be continuously transformed into one another and substituted for one another. I am not alone in thinking that there are serious problems with these ideas when it comes to the analysis of science. As agents, we humans seem to be importantly different from nonhuman agents like the weather, television sets, or particle accelerators. Certainly, for example, even the proponents of the actor-network seem to have little difficulty separating people and things when they write about science, as Yves Gingras (1994) has forcefully pointed out. More precisely, the idea that, say, human beings can be substituted for machines (and vice versa) seems to me a mistake. I find it hard to imagine any combination of naked human minds and bodies that could substitute for a telescope, never mind an electron microscope, or for a machine tool, or for an atom bomb (or for penicillin, heroin, . . .). Semiotically, these things can be made equivalent; in practice they are not.[23] But still, I believe that the actor-network is onto something in its extended symmetry, actually two things. One is that there exist important *parallels* between human and material agency, concerning both their repetitive quality and their temporal emergence; the other is that a constitutive *intertwining* exists between material and human agency. Or so my studies of practice convince me. I now want to explain these remarks, in a performative rather than a semiotic idiom.

23. The actor-network speaks of "delegating" human performances to machines; my suggestion is that the putative symmetry of this operation often breaks down when one tries to imagine delegating machinic functions back to humans.

When I have finished, most of the important elements of my understanding of practice will be in place, and I can pull them together in the next section.

We can start with scientific culture, as I defined it earlier, rather than with practice. Think of the field of machines that constitute the established material performativity of science at any given time. This machinic field does not exist in a human vacuum. Though the machines and instruments of science often display superhuman capacities, their performativity is nevertheless enveloped by the human realm. It is enveloped by human *practices* (recall the discussion of section 1.1)—by the gestures, skills, and whatever required to set machines in motion and to channel and exploit their power. Though it is not strictly about machines, one can think, for example, of Fleck's classic discussion of the history of the Wassermann reaction as a test for syphilis ([1935] 1979), in which he emphasizes the importance of the establishment of a community of practitioners competent to carry through the test. In the absence of such a community and its specific human performances, the test would not exist. Further, just as the performativity of machines is repetitive—my computer displays the same powers day after day—so is the human performativity that envelops them. The field of practices is routinized and disciplined, *machinelike,* as Collins puts it.[24] The proper performance of the Wassermann reaction requires adherence to a standardized sequence of gestures and manipulations; around machines, we act like machines. In these simple senses, then, there is, if not a perfect interchangeability, a very important degree of symmetry and interconnection between human agency and material agency: as respectively disciplined in practices and as captured in machines, they are both repetitive and machinelike and they collaborate in performances.

Now we can turn to practice, to the extension of scientific culture, and especially to the extension of the machinic field of science. I have already described this as involving a process of tuning, and the key point about tuning in the present context is that *it works both ways,* on human as well as nonhuman agency. Just as the material contours and performativity of new machines have to be found out in the real time of practice, so too do the human skills, gestures, and practices that will envelop them. It would make no sense, for example, to try to imagine

24. See Collins 1990, 1994; Collins and Kusch forthcoming a, forthcoming b. Collins speaks interchangeably of "machine-like" and "behaviour-specific" action; I prefer the former locution, for obvious reasons.

some community developing the skills implicit in the execution of the Wassermann reaction in the absence of the material technologies of the procedure. Gestures, skills, and so on—all of these aspects of disciplined human agency come together with the machines that they set in motion and exploit.[25] Here, then, we can note further degrees of symmetry between human and material agency. Just as material agency is temporally emergent in practice, so, necessarily, is disciplined human agency. And, furthermore, while the two are not continuously deformable into one another—one cannot substitute human beings for the reagents of the Wassermann reaction—they are intimately connected with one another, reciprocally and emergently defining and sustaining each other. Disciplined human agency and captured material agency are, as I say, constitutively intertwined; they are *interactively stabilized* (Pickering 1989c).[26]

There are other aspects of human agency besides machinelike practices that we will need to discuss as we go along; I will later talk about the scale and social relations of human actors and argue that these too are tuned and interactively stabilized in practice. For the moment, though, I want to discuss an aspect in which the symmetry between human and material agency appears to break down. I want to talk about *intentionality*—a term I use in an everyday sense to point to the fact that scientific practice is typically organized around *specific plans and goals*.[27] I find that I cannot make sense of the studies that follow without reference to the intentions of scientists, to their goals and plans, though I do not find it necessary to have insight into the intentions of things. One has to recognize that scientists usually work with some future destination in view, whereas it does not help at all to think about machines in the same way. Human intentionality, then, appears to have no coun-

25. This is Knorr-Cetina's point (1992, 1994) when she writes about the laboratory as the home and site of mutual reconfiguration of "enhanced nature" and "enhanced agents"—though the implicit contrast between "nature" and "agents" seems misplaced to me. Polanyi 1958 is an early classic on skill and "tacit knowledge" in science.

26. Or, as Callon and Latour (1992) put it (following Engels), coproduced, in what John Law (1987) calls a process of heterogeneous engineering.

27. Piet Hut has encouraged me to note that my usage of "intentionality" is not the technical usage found in phenomenological philosophy, where it evokes the idea that consciousness is always consciousness of something. On the other hand, there seem to be many points of contact between my overall analysis and Husserl 1970. The discussion that follows here, for example, could be taken as an elaboration of Husserl's remark that "[w]orld is the universal field into which all our acts, whether of experiencing, of knowing, or of outward action, are directed. From this field, or from objects in each case already given, come all affections, transforming themselves in each case into actions" (1970, 144).

terpart in the material realm. But still, I want to stress here the significant parallels and intertwinings between the intentional structure of human action and material agency. Especially I want to stress the temporal emergence of plans and goals and their transformability in encounters with material agency.[28]

We could start from the idea that perhaps a perfect symmetry does obtain between human and nonhuman agency, even if human intentionality is included in the picture. At some deep level, perhaps the vectors of human practice are just as temporally emergent from moment to moment and situation to situation as is material agency. Just as we do not know what this new machine will do, we do not know what other people, or even we ourselves, will do next. Think of the characters in J. G. Ballard's stories—always acting, but never with any vision of where their actions will take them. From this perspective, there would be an exact parallel between the intentional structure of human agency and material agency prior to any machinic capture: both are wild and undomesticated. In the studies that follow, however, human goals and purposes do not seem to be as wild as all that. They appear as already partially tamed, already on the way to being brought to heel by the cultures in which they are situated, as I can now explain.

In trying to understand the intentionality of scientific practice, it is important to continue to pay attention to time. Sometimes, at least, we humans live in time in a particular way. Unlike DNA double helixes or stereo systems, we construct goals that refer to presently nonexistent future states and then seek to bring them about.[29] We aim to build a new kind of machine, say, that we hope will display certain powers. This

28. Bruno Latour and Barbara Herrnstein Smith have criticized my views on intentionality, though I cannot discern any inconsistency between the position developed here and theirs. On Latour, see note 29 below; on Smith, see her words at the beginning of section 2. 3. Two further points are worth noting. First, the question of intentionality, as I have defined it, simply does not arise in synchronous studies of culture (including studies of practices). Second, the following argument states, and the studies discussed in part 1 exemplify, my conviction that the plans and goals of practice are not derivable from existing culture. No static account—of scientific reason as "problem-solving," or of "constraints" upon practice, or whatever—can explain why particular actors set off along particular vectors of cultural extension. Something more is needed, which my account seeks to supply.

29. Ashmore 1993 is a brave and amusing, but not very persuasive, attempt to attribute intentionality to a material agent, namely, a "catflap." The other way to preserve a symmetry between nonhuman and human agency is, of course, to deny intentionality to humans. This, I think, is John Law's strategy (1993) in describing human agents as "network effects," "performed" by organizational narratives or "myths" (note the return from

extended temporal sweep of human agency is, for me, a respect in which the symmetry between human and material agency breaks down, and without marking this difference I cannot make sense of the case studies that follow. But having marked it, more needs to be said. If one defines intentionality in terms of human plans and goals, the questions that arise concern the origin and substance of such goals, and here traditional studies of science already offer an answer. The goals of scientific practice are imaginatively transformed versions of its present. The future states of scientific culture at which practice aims are constructed from existing culture in a process of *modelling* (metaphor, analogy).[30] This, stated in a few words, is my basic idea of how existing culture predisciplines the extended temporality of human intentionality. I will exemplify and elaborate this idea later, but here I want just to make three comments that will connect it to my present theme of the parallels between and intertwining of human and material agency.

First, the predisciplining of intent by existing culture is only partial. Modelling is an open-ended process with no determinate destination. From a given model—say a particular functioning machine—an indefinite number of future variants can be constructed. Nothing about the model itself fixes which of them will figure as the goal for a particular passage of practice (Barnes 1982 expresses this point very well).[31] There is no algorithm that determines the vectors of cultural extension, which is as much as to say that the goals of scientific practice emerge in the

agency to textuality, paralleling the semiotic turn we have been discussing). It is worth noting that Latour often seems to have a pretty distinct notion of human intentionality. While the early image in Latour and Woolgar 1986 of an "agonistic" war of all against all—a general intention to dominate in battle—can be carried over symmetrically to nonhuman agents or to networks as wholes (Callon and Latour 1981), I do not think that the same can be said of Latour's later discussion of "translating interests" (1987, 108–21). This latter seems to me to be applicable only to intentional human agents acting on other intentional human agents. For a thoughtful review of sociological understandings of intentionality, see Lynch 1992c; for a survey of recent thinking in the social sciences on human and nonhuman agency, see the contributions to Ashmore, Wooffitt, and Harding 1994.

30. I take the idea of modelling from discussions in history and philosophy of science of the role of metaphor and analogy in theory development. I prefer to speak of modelling since I want to apply the idea to the material and social, as well as the conceptual, aspects of scientific culture, while metaphor and analogy are usually taken as having textual referents. Modelling, in Kuhnian terms, is developing an exemplar. For access to the relevant literature, see, for example, Barnes 1982; Bloor 1991; Gooding 1990; Hesse 1966; Knorr-Cetina 1981; Kuhn 1970; and Pickering 1981c, 1984b.

31. Such open-endedness in human agency seems to me a necessary counterpart to the emergent quality of material agency; it is what makes it possible to bring the two into relation with one another.

real time of practice. Existing culture, to appropriate Michel Foucault's phrase (1972), is literally the *surface of emergence* for the intentional structure of human agency. Even in the domain of intentionality, then, this temporally emergent quality constitutes an important parallel between human and material agency. Second, if the field of existing machines serves as a surface of emergence for the goals of scientific practice, then human intentions are bound up and intertwined (in many ways) with prior captures of material agency in the reciprocal tuning of machines and disciplined human performances. The world of intentionality is, then, constitutively engaged with the world of material agency, even if the one cannot be substituted for the other. And third, I can note that, as I conceive it, tuning can also transform the goals of scientific practice. Scientists do not simply fix their goals once and for all and stick to them, come what may.[32] In the struggles with material agency that I call tuning, plans and goals too are at stake and liable to revision. And thus the intentional character of human agency has a further aspect of temporal emergence, being reconfigured itself in the real time of practice, as well as a further aspect of intertwining with material agency, being reciprocally redefined with the contours of material agency in tuning.[33]

Thus my preliminary analysis of the intentional structure of human agency. I find myself forced to regard the latter as differing from nonhuman agency in its temporal structure, through its orientation to goals located in the future. But still I need to emphasize that such goals should be seen as *in the plane of practice* (to paraphrase my opening quotation from Deleuze and Parnet) rather than as controlling practice from without. Goals are temporally emergent from culture (including machines and their material performativities) and can themselves be transformed in, and as an integral part of, real-time practice, which includes sensitive encounters with material agency.[34]

32. Fuller (1992) thinks that scientists should be taken to task for this; to the contrary, I take it to be central to scientific creativity.

33. Collins seems to me to be the only writer within the SSK canon who is directly concerned with the temporal emergence of human agency. See Collins 1992 and my essay review, 1987. More generally, recognition of emergence in human agency aligns my position with symbolic interactionist, ethnomethodological, and pragmatist sociologies: Denzin 1992 surveys the history of symbolic interactionism up to the present, stressing its links with pragmatism and ethnomethodology; Lynch 1992a and 1993b are good entry points for ethnomethodological studies of science; for access to pragmatist studies of science and technology, see Star 1991b, 1992; and Fujimura 1992. Laurel Graham and John Law have long encouraged me to think about the relation between my studies of science and symbolic interactionism; I regret that I did not follow their suggestions earlier.

1.4 THE MANGLE OF PRACTICE

"Do not worry if you think it is dark," he said to me, "because I am going to light the light and then mangle it for diversion and also for scientific truth."

Flann O'Brien, *The Third Policeman*

As promised, most of the elements of my understanding of scientific practice have been laid out in picking my way through the chicken debate. Now I can sum them up and begin to elaborate upon them. My basic image of science is a performative one, in which the performances—the doings—of human and material agency come to the fore. Scientists are human agents in a field of material agency which they struggle to capture in machines. Further, human and material agency are reciprocally and emergently intertwined in this struggle. Their contours emerge in the temporality of practice and are definitional of and sustain one another. Existing culture constitutes the surface of emergence for the intentional structure of scientific practice, and such practice consists in the reciprocal tuning of human and material agency, tuning that can itself reconfigure human intentions. The upshot of this process is, on occasion, the reconfiguration and extension of scientific culture—the construction and interactive stabilization of new machines and the disciplined human performances and relations that accompany them.

This is the skeleton of my overall scheme. Now I want to expand it a little by returning once more to the idea of tuning. This will bring us to the mangle. I find tuning a perceptive metaphor, but in the detailed analysis of practice it helps to decompose it along the following lines. Tuning in goal-oriented practice takes the form, I think, of a *dance of agency.* As active, intentional beings, scientists tentatively construct some new machine. They then adopt a passive role, monitoring the performance of the machine to see whatever capture of material agency it might effect. Symmetrically, this period of human passivity is the period in which

34. It might help to clarify my position if I were to endorse some sentiments expressed by Lucy Suchman against an alternative view of human intentionality. She writes: "[O]ur actions, while systematic, are never planned in the strong sense that cognitive science would have it. Plans are best viewed as a weak resource for what is primarily *ad hoc* activity. . . Stated in advance, plans are necessarily vague, insofar as they must accommodate the unforeseeable contingencies of particular situations" (1987, ix). And, "rather than subsume the details of action under the study of plans, [we should think about how] plans are subsumed by the larger problem of situated action" (1987, 50). I agree.

material agency actively manifests itself. Does the machine perform as intended? Has an intended capture of agency been effected? Typically the answer is no, in which case the response is another reversal of roles: human agency is once more active in a revision of modelling vectors, followed by another bout of human passivity and material performance, and so on.[35] The dance of agency, seen asymmetrically from the human end, thus takes the form of a *dialectic of resistance and accommodation,* where resistance denotes the failure to achieve an intended capture of agency in practice, and accommodation an active human strategy of response to resistance, which can include revisions to goals and intentions as well as to the material form of the machine in question and to the human frame of gestures and social relations that surround it.[36]

The practical, goal-oriented and goal-revising dialectic of resistance

35. Fleck ([1935] 1979) is the inspiration for the active/passive distinction made here. Fleck also uses the metaphor of "tuning," giving it a specifically musical sense: "Wassermann heard the tune that hummed in his mind but was not audible to those not involved. He and his co-workers listened and 'tuned' their 'sets' until these became selective. The melody could then be heard even by unbiased persons who were not involved" (86). Later, Fleck socializes the musical metaphor: "The matching of all the five required reagents, so as to maximize the effect of the reactions and ensure that the results are as clear as possible, requires experience. Even quasi-*orchestral practice* is needed if, as is usual, the test is performed by a team" (97). Likewise, Lynch argues that to try to make sense of activities in a student laboratory without reference to the events that the students are witnessing through microscopes "would be analogous to analyzing an audiovisual recording of a symphony with the sound turned off" (1991b, 52). These musical figures can be understood as part of an attempt to escape from traditional visual/representational understandings of science, but they remain, in the end, asymmetric: the metaphor should be dancing (with partners), not listening or even performing. As usual, Latour is more surefooted: "An experiment *shifts out* action from one frame of reference to another. Who is acting in this experiment? Pasteur *and* his yeast. More exactly, Pasteur acts *so that* the yeast acts alone. . . he creates a scene in which he does not have to create anything. He develops gestures, glassware, protocols, so that the entity, once shifted out, becomes automatic and autonomous. . . Who is doing the acting in the new medium of culture? *Pasteur,* since he sprinkles, and boils, and filters, and sees. *The lactic acid yeast,* since it grows fast, uses up its food, gains in power . . . , and enters into competition with other similar beings growing like plants in the same soil. If I ignore Pasteur's work, I fall into the pitfalls of realism . . . if we ignore the lactic acid . . . [w]e fall into the other pit . . . of social constructivism, forced to ignore the role of nonhumans . . . We do not have to choose between two accounts of scientific work, since this very scientific work aims at building a scene in which scientists do not do any work" (1992b, 141–44). (These sentiments do not seem very semiotic to me.)

36. I thank Yves Gingras for pointing out the close relation between what I call the dialectic of resistance and accommodation and Jean Piaget's account (1985) of intellectual development. "Accommodation" is one of Piaget's key terms, often understood in respect of "resistances" and "obstacles," and his notion of "equilibration" is close to my idea of interactive stabilization. Piaget's empirical interests are different from mine, and his ideas are elaborated more in a representational than in a performative idiom, but it would be

and accommodation is, as far as I can make out, a general feature of scientific practice. And it is, in the first instance, what I call *the mangle of practice,* or just the mangle. I find "mangle" a convenient and suggestive shorthand for the dialectic because, for me, it conjures up the image of the unpredictable transformations worked upon whatever gets fed into the old-fashioned device of the same name used to squeeze the water out of the washing. It draws attention to the emergently intertwined delineation and reconfiguration of machinic captures and human intentions, practices, and so on. The word "mangle" can also be used appropriately in other ways, for instance as a verb. Thus I say that the contours of material and social agency are mangled in practice, meaning emergently transformed and delineated in the dialectic of resistance and accommodation.[37] In a broader sense, though, throughout the book I will take the mangle to refer not just to this dialectic and the transformations it effects, but to the overall scheme, as I just called it, to the overarching image of practice that encompasses the dialectic—to the worldview or metaphysics, if you like, which sees science as an evolving field of human and material agencies reciprocally engaged in a play of resistance and accommodation in which the former seeks to capture the latter.

Thus the mangle, which in its several senses is the subject of all that follows. To get much further, we really need to turn to some examples, but before we do, one last remark might help. I do not think that the workings of the mangle are hard to grasp in any particular instance, but two aspects of the overall analysis are. One is the concept of temporal

---

interesting to work through the extension of his scheme to scientific practice. In Piaget's terms (1985, 17), my concern is with "homeorhesis" rather than with "homeostasis."

37. If pressed too hard, the mangle metaphor quickly breaks down. A real mangle leaves the list of clothing unchanged—"shirts in, shirts out"—which is too conservative an image for the constructive aspect of scientific practice. "Mangling" also carries connotations of mutilation and dismemberment—"my teddy bear was terminally mangled in a traffic accident"—which carry one directly away from this constructive aspect. There is little to be done about this; I can think of no more appropriate word; one has simply to try to take the metaphor seriously enough but not too seriously. (Alternatively, one could think of the productive and transformative magic mangle in Flann O'Brien's *The Third Policeman.*) I thank, among others, Mike Lynch, Ted O'Leary, and Allan Megill for warning me of potential difficulties with mangle talk and encouraging me to explain it more fully. (It turns out that there are even disputes about what a mangle is as domestic technology. Natalie Davis and Frederick Suppe tell me that their mothers' mangles were devices to speed up the ironing. This might suggest that the meaning of mangle differs between England, where I learned the term, and the US, but Richard Burian offered to take me into the depths of West Virginia and show me mangles, in my sense, still in use.)

emergence. What emergence entails is exemplified in the earlier brief discussion of material agency. In advance, we have no idea what precise collection of parts will constitute a working machine, nor do we have any idea of what its precise powers will be. There is no thread in the present that we can hang onto which determines the outcome of cultural extension. We just have to find out, in practice, by passing through the mangle, how the next capture of material agency is to be made and what it will look like. Captures and their properties in this sense *just happen.* This is my basic sense of emergence, a sense of brute chance, happening in time—and it is offensive to some deeply ingrained patterns of thought. The latter look for explanations—and the closer to the causal, mechanical explanations of classical physics the better—while it seems to me that in the analysis of real-time practice, in certain respects at least, none can be given. I can do nothing about this, but it is best to be clear on this point from the start. The world of the mangle lacks the comforting causality of traditional physics or engineering, or of sociology for that matter, with its traditional repertoire of enduring causes (interests) and constraints. I must add, though, that in my analysis brute contingency is constitutively interwoven into *a pattern that we can grasp* and understand, and which, as far as I am concerned, does explain what is going on. That explanation is what my analysis of goal formation as modelling, the dance of agency, and the dialectic of resistance and accommodation is intended to accomplish. The pattern repeats itself endlessly, but the substance of resistance and accommodation continually emerges unpredictably within it.[38]

38. At stake here is the question of what explanation and understanding consist in, a topic of interesting debate in contemporary philosophy. Mangle-ish explanation of scientific practice has much the same explanatory character as that ascribed to chaos theory in the natural sciences by Stephen Kellert: in contrast to traditional scientific explanations, "chaos theory does not provide predictions of quantitative detail but of qualitative features; it does not reveal hidden causal processes but displays geometric mechanisms; and it does not yield lawlike necessity but reveals patterns. . . Chaos theory takes up [an] emphasis on finding patterns and connections, while jettisoning the requirement that the patterns must yield necessity in a detailed and deterministic sense" (1993, 105, 112). One might argue that this just shows that chaos theory (and the mangle) explains nothing, but Kellert notes on behalf of chaos theory—as I do later for the mangle—that there are many natural phenomena on which traditional explanatory schemata find no purchase: "an account of scientific understanding as the disclosure of hidden causal processes is not only inadequate for the biological and social sciences . . . but it is inadequate for the physics of nonlinear systems as well" (105). Kellert also (113) draws interestingly upon Evelyn Fox Keller's discussion (1985) of "order" as a generalization of traditional notions of "law" in science. Latour (1988a) makes some interesting and perceptive observations and argu-

Now for the other difficult aspect of the mangle. We can return for a moment to the chicken debate. Collins and Yearley operate on a terrain that I find much more familiar than Callon and Latour's (one needs to read Callon and Latour's essay to appreciate this point fully), and here we run into the problem of the "posts." Collins and Yearley's arguments are familiar, I think, because they are modernist, humanist, dualist; Callon and Latour's are less familiar because they are postmodern (though Latour [1990, 1993b] does not like the term), posthuman, postdualist (if that is a word).[39] To explain what is going on, I can dwell on the most perspicuous of these distinctions, humanist/posthumanist. Collins and Yearley's position, like SSK in general, is grounded in the traditional humanism of sociology as a discipline, inasmuch as it takes the human subject to be the center of the action. Whenever SSK detects a tendency for the action to be located elsewhere—a decentering of the human subject—it undertakes the police action already discussed. Any trace of nonhuman agency is immediately recuperated by translating it into an account of nonhuman agency that is attributed straight back to human subjects. Or, as we have seen, it is admitted that one can speak naively about nonhuman agency, but if one wants to do that, according to Collins and Yearley, one has to speak in a possibly even more familiar and antihumanist idiom, that of the scientists and engineers. Their discourse is a pure one of material agencies from which human agency is quite absent. (Thus Collins and Yearley's dualism.)

These humanist and antihumanist discourses run deeply through everyday thought, though they do not exhaust it. They are also the very stuff from which the traditional academic disciplines are created and that holds them apart (I come back to this in section 7.1). To be a traditional sociologist *is* to be a humanist; to be a physicist *is* to be an antihumanist (in technical practice, I mean). But the mangle, like the actor-network approach, corrodes the distinctions these discourses and disciplines enforce. It is not just that the mangle and the actor-network multiply sites of agency, trying to keep both human and nonhuman agency in view at the same time. Rather, in their somewhat different ways, the mangle and the actor-network insist on the constitutive

ments about explanation in the context of actor-network theory, though he couches them in the representational/semiotic idiom that I seek to avoid.

39. People usually say monist instead of postdualist; in the light of the discussions of cultural multiplicity in section 1.5, though, I think a better word would be "multiplist." My analysis is on the other side of two from one.

intertwining and reciprocal interdefinition of human and material agency. The performative idiom that I seek to develop thus subverts the black-and-white distinctions of humanism/antihumanism and moves into a *posthumanist* space, a space in which the human actors are still there but now inextricably entangled with the nonhuman, no longer at the center of the action and calling the shots. The world makes us in one and the same process as we make the world. This posthumanism is the second aspect of the mangle that thought tends to bounce off and even recoil from. Again, I make no apologies for it. It seems to me to be how things are in practice, an interesting but difficult observation that needs explicitly to be made, and the mangle helps to keep it in view.[40] To speak for myself, I am not sure whether the temporal emergence of the mangle or its posthumanism is the harder to take. Perhaps it is the entanglement of the two, since my way of understanding the interconnection of human and material agency is essentially temporal.

In any event, to get clear on what is at stake I return to these themes of emergence and posthumanism as the book progresses. Reviewing the studies that follow, I try to be as clear as I can on the details of just how they exemplify temporal emergence and the decentering of the human subject in the extension of culture. From time to time, I also explicitly contrast my analysis of this process with traditional—humanist and nonemergent—accounts.[41] My prime example of the latter is the approach I have already discussed most—SSK. And some explanation is needed for this expository tactic and choice of foil. I have tried, in general, to write this book constructively, laying out my own understanding of scientific practice as best I can. I have, however, come to the conclusion that the general trend of my arguments is best grasped by the kind of periodic confrontation with the more traditional views just men-

40. I finally grasped the significance of the humanist/posthumanist distinction in conversations with John Law and in reading his recent work (Law 1993, for example). Barbara Herrnstein Smith also helped me by pointing out a "residual humanism" in an earlier essay of mine (and the residue is arguably still there in the above discussion of human intentionality). Beyond the actor-network literature, I have found the works of Gilles Deleuze and Félix Guattari (especially Deleuze and Guattari 1987) and Michel Foucault very useful in thinking about the decentering of the human subject. Foucault (1972) offers a general elaboration of the themes of temporal emergence and the displacement of the human subject, though how he connects them is not clear to me. I link them via the preceding discussion of agency, a topic on which Foucault displays a principled reluctance to speak.

41. From now on I use "traditional" as shorthand for humanist, nonemergent, and representationalist approaches in science studies.

tioned. Among the latter, in most of these confrontations I single out SSK for three reasons. First, because I find it by far the most plausible and interesting nonemergent humanist account of science that is available (much more plausible and interesting than, say, accounts couched in terms of epistemic rule following—though I do discuss those, too, in section 6.4). Second, because I suspect that SSK has *succeeded too well.* I mentioned earlier the invaluable role that SSK has played in widening our conceptions of science as an object of study, but it seems to me that much of the work currently being done in science studies has remained stuck in the place where SSK left it in the early 1980s. Perhaps this is not the case for specialists in the sociology of science. Even the later work of some its founders, like Steven Shapin, often bears little resemblance to SSK's prototypical form, which is why I tend to refer to "early" or "canonical" SSK in subsequent comparisons and confrontations. But many historians and philosophers who have become sensitized to science's social dimensions continue to treat SSK as if it were the last word on the subject (witness the current fashion for multidisciplinary eclecticism, discussed in section 7.1, or the dominance of interest explanations in STS). In taking SSK as a primary exemplar of nonemergent humanism, then, my aim is therapeutic: I would like to help end the sleep induced by this particular spell.[42] My third reason for discussing SSK is a constructive reformulation of the second. SSK tends to treat the social as a nonemergent causal explanation of cultural extension in science. In contrast, my studies convince me that the social is bound up with science and itself subject to mangling in scientific practice. Unlike SSK, then, I am directly interested in processes of social transformation. I want to analyze them, and comparisons with SSK in particular can thus help me to foreground what is at stake in this aspect of my analysis.

1.5 MORE ON THE MANGLE

There are angles on the mangle that remain to be mentioned, and the easiest way to get at them is to look ahead. The rest of the book is divided into two parts. Part 1, comprising chapters 2 to 5, revolves around

42. An alternative rhetorical strategy would be to describe the mangle (together with other nontraditional approaches to science studies) as a part of what SSK has become. Trevor Pinch, for example, has tried to persuade me of such a broad construal of SSK. Even following this line of thought, however, I think that it would remain important to make it explicit that SSK has taken an emergent and posthumanist turn away from its origins and canonical works. I also think that this construal tortures the words "sociology

four case studies intended to exemplify and elaborate upon aspects of the mangle. Part 2, chapters 6 and 7, follows up implications and leads in philosophy, social theory, and historiography of science.

As one would expect, chapter 2 is about the machinic base of science. It focuses on the construction of an instrument, the bubble chamber as a tool for elementary-particle physics. Following Peter Galison, I trace out the evolution of the bubble chamber from its first conception to its realization as a working device, and analyze its historical development as an exemplification of the mangle. This chapter thus puts some flesh on my ideas concerning the capture of material agency, modelling, and intentionality, the intertwined emergence of material and human agency in the dialectic of resistance and accommodation, and so on. As far as human agency is concerned, I talk first about the mangling of its intentional structure as discussed above. I then turn to a different aspect not so far discussed, emphasizing the mangling of the *social contours* of human agency—the transformation and interactive stabilization in practice of both the scale of social actors and their relations to other such actors. And I seek to clarify what is at stake in my analysis through a comparison between my account of human agency and traditional accounts framed in terms of enduring causes and constraints, emphasizing the emergent and posthuman displacements effected by the mangle. I can note here that appeals to notions of nonemergent constraint seem to be running rampant at present, even among authors otherwise opposed to traditional philosophy and social theory. I hope that my analysis can help stem the flood, and I continue the argument in chapter 5 (against the related notion of "limits") and in chapter 6 (in my discussion of relativism). Besides topics of agency, chapter 2 also begins my analysis of how knowledge can be understood within the performative idiom, arguing that representations of how the bubble chamber functioned were themselves mangled in practice.

Like chapter 2, chapter 3 is concerned with the machinic base of science and captures of agency, but there I am more directly interested in how knowledge is threaded into the machinic field of science—with how one can think representation in the performative idiom, with, espe-

of scientific knowledge." The move to the performative idiom strips knowledge of the special privilege that the *K* of "SSK" accords it; likewise the antidisciplinary tendency of the mangle (see above and section 7.1) argues against the disciplinary allegiance of the first *S*. The second *S* might stand, though I argue in chapter 5 and sections 7.3 and 7.4 that on my analysis science bleeds right into technology, the factory, and society.

cially, the production of scientific facts. The empirical study is of the construction of an instrument designed to investigate the existence of tiny particles called quarks, and of the use of that instrument to generate empirical knowledge of immediate theoretical relevance. An aspect of science that I mentioned earlier in this chapter but have since left undeveloped becomes crucial here, namely, its cultural multiplicity and heterogeneity. I argue that scientific practice is, in general, organized around the making (and breaking) of *associations* or alignments between multiple cultural elements, and that fact production, in particular, depends on making associations between the heterogeneous realms of machinic performance and representation, in a process that entails the emergent mangling and interactive stabilization of both. Articulated knowledge and machinic performances are reciprocally tuned to one another, I suggest, in a process that involves the artful *framing* of already captured material agency. At the end of the chapter, I draw upon other studies of science, first to elaborate an overall image of scientific knowledge as sustained in extended representational chains terminating in captures and framings of material agency, and then to argue for the mangling and interactive stabilization of the disciplined practices that envelop, sustain, and are sustained by such chains.

Chapter 4 continues the attempt to understand representation in the performative idiom, and is entirely about conceptual practice, the business of extending the conceptual apparatus of science. Its underlying concern is with how conceptual practice can have a shape: why is it not completely arbitrary or subjective? This question arises within my analysis because conceptual practice proceeds in the absence of any immediate engagement with the material world that could help constitute the kinds of dialectic of resistance and accommodation that the two preceding chapters dwell upon. So what, if anything, do scientists have to struggle with in the realm of concepts and thought? What lies between them and their goals? My answer to such questions depends upon seeing conceptual practice as again seeking to create alignments between multiple cultural elements, and locates the possibility of resistance in this process not now in material agency but in what I call *disciplinary agency*—the sedimented, socially sustained routines of human agency that accompany conceptual structures as well as machines. I show how disciplinary agency can play an analogous role in conceptual practice to that of material agency in material practice. And I show, too, how disciplinary agency is itself mangled in conceptual practice. I should note that my example of conceptual practice comes from mathematics rather than science proper: it concerns Sir William Rowan Hamilton's formulation of the "quaternion" system in 1843. I am happy that my analysis of the mangle can be extended to mathematics, but I hope that this chapter will also be read as an indication of how conceptual practice in science, including the realm of scientific theory, might likewise be conceived. The chapter ends with a summary of the analysis developed in chapters 2 to 4, followed by a postscript in which I discuss David Bloor's SSK-style analysis of Hamilton's work. Bloor suggests that Hamilton's metaphysical orientation was dependent upon his social position and interests; my suggestion instead is that Hamilton's metaphysics was emergently and posthumanly mangled in his technical practice.

Chapter 5 moves us into a rather different space, so before I come to it I want to make a general comment on chapters 2 to 4. These chapters constitute the empirical body of my performative understanding of science. They run upward from the construction and use of machines and instruments, through the production of facts and their engagement with theory, to the quasi-autonomous dynamics of thought and representation. That sounds like quite a lot, but, having worked through and reflected upon these studies I am conscious of what a sparse set of examples they are in relation to what I have come to understand as the immense complexity of scientific culture and practice. More particularly, it is clear that my terms of engagement with practice are tuned to the physical/mathematical sciences (and probably many branches of engineering) rather than to other sciences. The word "machine" comes less and less naturally as one moves from physics even to chemistry, let alone to fields like biology or the social sciences. I confess, though, that I am not very disturbed by this. As the writings of the actor-network amply demonstrate, ideas about nonhuman agency are more readily graspable in respect of the biological than the physical sciences, and the work of Michel Foucault points the way to a parallel analysis of the human sciences. One can see well enough, then, how my account of scientific practice might be extended beyond its present domain of exemplification (which is not to say that detailed explorations would not be illuminating).[43]

43. For example, I find many suggestive connections between my analysis of the mangle and Foucault's discussion (1979) of the prison. In my terms, the prison can be seen (like other institutions) as a site for capturing, channeling, and framing the agency of a particular segment of the human population, and much of Foucault's analysis relates to a kind of institutional tuning or mangling, in which the social sciences have themselves emerged and continue to be mangled. More generally, I think that my discussion of the

Now for chapter 5. This breaks the sequence established by chapters 2 to 4. It is not about science per se. It is about technology and the workplace, about industry, production, capital, and labor relations. It is, to be specific, about the introduction of the new technology of numerically controlled machine tools into the factory. My principal reason for its inclusion is that it shows that my analysis of the mangle can be put to work outside the realm of science. As in chapters 2 and 3, a notion of capturing material agency is at the heart of chapter 5, as is that of the mangling of the social—here meaning the formalized and disciplined roles and relations of workers and management. The point that I want to emphasize in this chapter is, then, that a performative analysis like the mangle runs very smoothly and directly between the worlds of science, technology, and work: it promises an integrated vision of STS. A single example cannot, of course, make the case, but the overall conclusion is something that the actor-network has already reached (see, for example, Latour 1987), and I want to show how we can arrive at it in the performative idiom. A subsidiary objective of this chapter is to continue the confrontation with traditional understandings of practice. I take my empirical information from David Noble's *Forces of Production,* a book in which Noble offers a quite different interpretation from mine of the events at issue. His is SSK-like, a traditional, nonemergent, and humanist interpretation based upon enduring social actors and their interests, in which a notion of enduring limits to practice plays a crucial role; mine is emergent and posthumanist, foregrounding the transformations in the social that are part of the encounter with material agency. A critical comparison of our two approaches serves, therefore, to highlight what is more generally at stake in the move to the mangle.

The two chapters that make up part 2 of the book take those of part 1 as their point of departure. Chapter 6 is especially concerned with appreciations of scientific knowledge that follow from the mangle. It begins by trying to get to grips with the problematic of realism, which I take to center on the nature of the purchase our knowledge has on the world and vice versa. I argue that the mangle offers us a realistic appreciation of scientific knowledge inasmuch as it demonstrates the nontrivi-

mangling of representational chains in chapter 3 could readily be generalized to the relays and alignments in chains of power/knowledge that Foucault has thematized. In this connection, see the Foucauldian analyses of developments in the social sciences, management, and accountancy by Hopwood (1987), Miller (1992), Miller and O'Leary (1994), Rose (1990), and Rose and Miller (1992).

ality of the construction of representational chains terminating in captures and framings of material agency. I call this the *pragmatic realism* of the mangle, to distinguish it from the correspondence realism that defines the traditional realist-antirealist debate in philosophy. I argue that pragmatic realism is not, in the first instance, a position in the debate over whether representations correspond to nature, but that it subverts that debate. Once one appreciates the pragmatic realist engagement of knowledge with the world, questions of correspondence no longer seem pressing. One can, however, make trouble for correspondence realism on the basis of its pragmatic namesake. I argue that it is consistent with pragmatic realism to suppose that the world will support an indefinitely diverse set of ontologies and bodies of knowledge, each terminating in its own particular field of machines, and I offer historical exemplification of this possibility. At stake here, then, is an idea of *machinic incommensurability* that cannot be formulated in the representational idiom but that makes it hard to suppose that any particular scientific ontology exists in a special relation of correspondence to nature itself.

I return to the discussion of incommensurability in chapter 7, but the rest of chapter 6 addresses issues concerning objectivism about scientific knowledge and alternatives to it. I begin by defining the sense in which the mangle is an analysis of the objectivity of scientific knowledge. My argument is that maneuvers in the field of material and disciplinary agency immediately delocalize and confer a degree of objectivity on the products of scientific practice even at the individual level. On my analysis of practice, it is far from the case in science that "anything goes." However, as one would expect from my comments on incommensurability, the objectivity of the mangle is itself consistent with a certain relativity of scientific knowledge, a dependence of knowledge upon the conditions of its making, evaluation, and use. Scientific knowledge, on my account, is both objective and relative. Care needs to be taken, however, with these terms. I contrast my analysis of objectivity with the traditional philosophical one that looks to nonemergent and humanist rules of reason to guarantee the objectivity of theory choice. I argue that the mangle subverts this picture, and that whatever rules one is inclined to see as operative in science have themselves to be seen as in the plane of practice, and as liable to emergent mangling in one and the same process as all of the other heterogeneous elements of scientific culture. I then make a similar argument concerning traditional accounts of the relativity of science in terms of nonemergent interests or constraints (or limits) that have been held to specify the dependence of knowledge upon the social. The thrust of the studies in part 1 is that whatever characteristics of the social one cares to identify need, again, to be seen as subject to mangling, as themselves remade in practice. On my account, then, there is no enduring substantive link to hang onto in grasping the relation between what culture is and what it is to become—neither epistemic rules nor properties of the social fulfill this function. There is only the mangle, and what it cares emergently to grind out. And this, in turn, speaks of an irremediable historicity of scientific knowledge (and culture in general): what counts as knowledge now is a function of the specific historical trajectory that practice has traced out in the past. To emphasize this point, the chapter concludes by continuing the discussion of the quark-search experiments of chapter 3 into a controversy that came briefly to surround them. Controversy is a classic site of articulation for SSK, which argues that divergent accounts of nature call for distinctively sociological explanation. In contrast, I suggest that nothing is needed to explain the divergence at issue beyond the mangle and the real-time contingencies of its working. The conclusion of this phase of the discussion is, then, that scientific knowledge is objective, relative, and historical, all at once.

Chapter 7 turns from the study of scientific practice to the practice of science studies, and addresses two main topics. One concerns possibilities for synthesis among the fields that make up science studies. I note that traditional explanatory structures lead directly into the well-known divergences and battles between disciplinary (philosophical, sociological, and so on) accounts of science and that, in their own terms, these divergences can only be overcome by adding up competing accounts in a strategy that I call multidisciplinary eclecticism. The mangle can instead contribute to, and offer a rationale for, a new and genuine antidisciplinary synthesis—deflating traditional schemes by displacing disciplinary variables (rules, interests, and so on) onto the same plane as technical scientific practice itself. I then turn to what must be a central component of the new synthesis: cultural studies of science, as defined in section 1.1. I argue that cultural studies and the mangle should be seen as orthogonal but complementary approaches to science studies, each contributing to a common understanding of science along its own axis; and I try to show how the new synthesis in science studies thus delineated opens out into a broader antidisciplinary synthesis of which the field of cultural studies per se is a central marker.

The second principal topic of chapter 7 is the question of how to get from the micro to the macro. Noting that the studies of part 1 relate to the practice of individuals or relatively small groups, I enquire here into the implications of the mangle for how we might think about practice at a more aggregated social level. Having had my say on topics in philosophy and social theory earlier in the book, at this stage I concentrate on historiography, on how, especially, we might write history in the large. I suggest that the move from the representational to the performative idiom points toward an integrated historiography of STS in which, to put it iconically, the Industrial Revolution takes over the special place occupied by the Scientific Revolution in traditional history of science. I then exemplify the possibility of getting to grips with macrohistory as *macromangling* in a discussion of the intersection of scientific and military enterprise in World War II. My argument here is that my analysis is *scale invariant:* that practice at any level of aggregation can be understood in terms of emergent and posthuman mangling. I further note that the study of macromangling promotes a desirable kind of double vision, a historiographic sensitivity both to the mangle-ish character of history in general and, at the same time, to the specificity of particular historical developments and formations.

Chapter 7 concludes with two postscripts—optional extras. In the body of the book, I take for granted certain standard but nevertheless circumscribed conceptions of what people and machines can do. In the first postscript, I note that outside industrialized societies quite nonstandard ideas of human and material agency are common, and I argue, first, that my analysis could readily be extended to nonstandard agency and, second, that a recognition of nonstandard agency points to a very radical form of incommensurability, an incommensurability of *powers.* The second postscript develops an idea of the mangle as a Theory of Everything. If one deletes the specifically human features of my analysis—in particular, the discussion of human intentionality broached earlier—I cannot see why it should not apply to the world itself: to chemical, biological, geological, cosmological systems, and whatever. The stumbling block is, of course, the rival nonemergent accounts offered by the existing sciences, and the book ends with some thoughts on how to get around them.

PART ONE

Instantiations

TWO

Machines

Building the Bubble Chamber

In chapter 1, I laid out my basic understanding of the mangle of practice in general terms. This chapter begins the task of exemplification and elaboration. It centers on a study of the extension of the machinic field of science, specifically of the development of the bubble chamber as an instrument for experimental research in elementary-particle physics.[1] Drawing primarily on Peter Galison's detailed account (1985), I begin with a historical narrative, concentrating on the work of Donald Glaser, the chamber's inventor, exemplifying my key concepts of resistance and accommodation—the mangle, in its restricted sense—and emphasizing the goal-oriented nature of Glaser's practice. I then offer a commentary organized around my themes of agency and emergence. I suggest that we should see the chamber as a locus of nonhuman agency, and I argue that its material contours, its powers, and accounts of its functioning (scientific knowledge) were emergently produced in a real-time dialectic of resistance and accommodation—they were, as I say, mangled. Turning to human agency, I analyze the intentional structure of Glaser's practice in terms of modelling, and argue that Glaser's plans and goals were

1. Several reasons recommend this particular example. The choice to focus on a scientific instrument is indicated by my present concern with material agency, and this is a historically significant instrument, central to two Nobel prizes (see below). As will also become evident, the history of the bubble chamber has an interesting social dimension that many similar histories lack. This social dimension, however, is not so rich or elaborate that it dominates the story, which is as it should be if the structure of the posthumanist displacement of the mangle is to be clearly expressed. Finally, there exists an excellent account of the history of the bubble chamber (Galison 1985) on which I can draw to establish my central points without myself telling the story in detail.

likewise emergently mangled. I further note that this mangling extended to the social contours—the scale and social relations—of human agency (I incorporate additional material on Luis Alvarez's practice around the bubble chamber to elaborate this point). Not just the vectors of human agency can be transformed in practice, then: the unit of analysis can change, too. Finally, I compare my account of human agency with traditional humanist accounts in a way that highlights the posthumanist intertwining of human and material agencies in the mangle.

2.1 BUILDING THE BUBBLE CHAMBER

The bubble chamber was invented in the early 1950s and became the principal tool of experimental elementary-particle physics in the 1960s and 1970s. Understanding its genesis and development requires some technical background. A typical particle-physics experiment has three elements. First, there is a beam of particles—protons, electrons, or whatever. This beam can be derived from natural sources, such as the flux of cosmic rays which rains sporadically upon the earth, or it can be artificially produced in a particle accelerator. Second, there is a target—a chunk of matter that the beam impinges on. Particles in the beam interact with atomic nuclei in the target, scattering—changing energy and direction—and often producing new particles. Third, there is a detector, which registers the passage of the scattered and produced particles in a form suitable for subsequent analysis.

The instruments that we need to think about in this chapter are cloud chambers and bubble chambers, and since their working principle is similar, a description of the former will suffice for the moment.[2] A cloud chamber is basically a tank full of vapor held under pressure, which doubles as both target and detector. Particle beams impinge on the vapor and interact and scatter there; when the pressure is released, the vapor begins to condense, and small droplets of liquid form first as strings marking the trajectories of any charged particles that have recently passed through it. These "tracks" are photographed, as permanent records of any particle interactions or "events" that occurred

2. For more on the basic strategies of elementary-particle experiment in the period at issue, see Pickering 1984b, chap. 2; on the early history of the cloud chamber, see Galison and Assmus 1989.

within the chamber. Now to history (page number citations are to Galison's essay).

In the early 1950s, a problem was widely recognized in the physics of the so-called strange particles. These had been discovered in cosmic-ray experiments using cloud chambers, but it was proving very hard to accumulate data on them. Strange-particle events seemed to be very rare. At this point, Donald Glaser, then beginning his career as an assistant professor at the University of Michigan, set himself a new goal. He wanted to construct some new kind of detector, like the cloud chambers that he had worked with as a graduate student, but containing some denser working substance. His reasoning was simple: event rates are proportional to the mass of the target for a given beam intensity, so if he could work with a denser medium, he would stand more chance of finding the strange-particle events of interest. He began to investigate a range of techniques using liquids and solids that would, he hoped, register particle tracks like those produced in cloud chambers, but these failed, one after the other.[3] None of them produced anything like a particle track. These failures constituted, to introduce one of my key terms, a sequence of *resistances* for Glaser, where by "resistance" I denote the occurrence of a block on the path to some goal. I should emphasize that I use "resistance" in just this sense of a practical obstacle, and I do not mean it to refer to whatever account scientists might offer of the source of such obstacles. More on such accounts later; first I want to describe Glaser's responses to such resistances as *accommodations:* in the face of each resistance he devised some other tentative approach toward his goal of a high-density detector that might, he hoped, circumvent the obstacles that he had already encountered. In the early trials, these accommodations took the simple form of moving from the exploration of one working substance and technique to the next. His practice took the form, then, of a dialectic of resistance and accommodation, which shifted him through the space of all of the potential new detector arrangements that he could think of. This dialectic is what, in its restricted sense, I call the mangle of practice.

Glaser's practice reached a temporary resting place in 1952 when he

3. Galison (317) mentions attempts to record polymerization reactions in liquids and to develop track-sensitive Geiger counters as well as "diffusion" cloud chambers (these last improving event rates by being continuously sensitive rather than by having higher density).

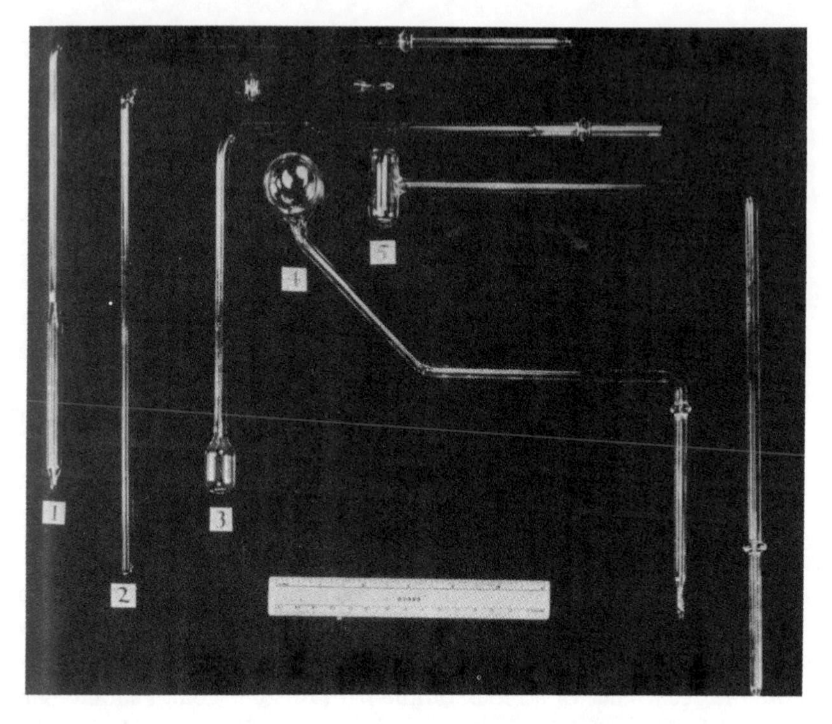

Fig. 2.1. Glaser's prototype chambers. The bulbous termini contain the volumes where tracks were observed. Reprinted, by permission, from Glaser 1964, 535, fig. 2. © The Nobel Foundation 1961.

built the first prototypes of a new detector that worked—the bubble chamber. Its operating principle was like that of the cloud chamber, but instead of being filled with a vapor it was filled with a superheated liquid held under pressure. When the pressure was released, boiling began, and small bubbles (instead of droplets) formed along the tracks of particles and could be photographed (figs. 2.1 and 2.2). The main thing was that the liquid filling of the bubble chamber was much denser than the vapor used in cloud chambers, and thus the former held out the promise of the higher event rates that defined Glaser's goal.

Glaser made his work public in early 1953, at which time several individuals and groups quickly set to work to develop the chamber into a practical instrument, most notably Glaser himself in Michigan, Luis Alvarez at Berkeley, and a group at the University of Chicago. (Glaser and Alvarez were awarded the Nobel Prize in physics for this work, in 1960 and 1968, respectively.) A point to note for future reference is that

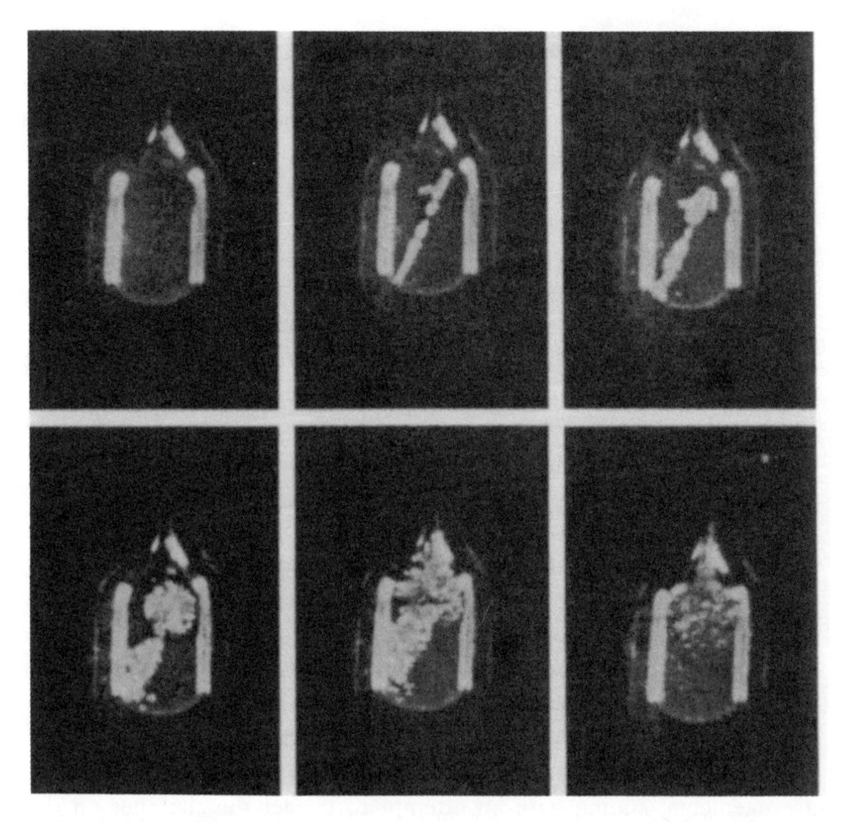

Fig. 2.2. Photographs of bubble tracks taken at increasing time intervals (0, 1/3, 1, 22, 71, and 250 milliseconds) in Glaser's prototype chamber. The well-defined track that has formed in the second photograph blurs as the bubbles that constitute it grow. Reprinted, by permission, from Glaser 1964, 536, fig. 3. © The Nobel Foundation 1961.

quite different goals were constructed around the bubble chamber at the three locations. In this respect, the work of Alvarez's group was the most impressive, establishing a basis for the big-science approach to particle-physics experiment that came increasingly to dominate the field in the 1960s. I will stay largely with Glaser, however, since his work offers a clear and simple exemplification of the mangle. I will make comparisons with Alvarez's work whenever they are useful.[4]

4. Galison 1985 gives an extensive account of Alvarez's bubble-chamber work; Pickering 1990b is an analysis of Alvarez's practice along the lines laid out here but lacking the terminology of the mangle.

Prior to his invention of the bubble chamber, Glaser had trained and worked as a cosmic-ray physicist. He was used, that is, to doing particle physics using naturally occurring cosmic rays rather than beams artificially produced in accelerators. And, after construction of the prototypes, his goal became that of inserting his new detector into his existing specialty. Here another resistance was apparent. Since cosmic rays arrive at the surface of the earth erratically, there was little chance of detecting interesting cosmic-ray events by expanding a bubble chamber at random intervals: the odds were high that nothing would be happening at the instant chosen. This problem was already familiar to physicists working with cloud chambers, and the established solution was to use a different kind of detector as a "trigger." A small electronic detector would be rigged up to register the passage of cosmic rays, and its output would be used to initiate the expansion of the chamber. In this way, photographs would be taken only when there was a good chance of finding interesting events. In the extension of his prototype chambers, Glaser adopted this triggering strategy—and failed. He found that when he wired up his bubble chamber to an electronic trigger it did not to produce any tracks. Once more a resistance had surfaced on the path to his intended goal, and once more a sequence of attempted accommodations followed. The material form of the bubble chamber was mangled in this process, as it was attached to a whole series of different triggering arrangements, ending with an attempt to trigger the chamber on the sonic plink that accompanied initial boiling.[5] None of these material transformations worked, and Glaser's next accommodation was more drastic.

Glaser's response to the continuing failure of his attempts to trigger his chamber on cosmic rays was, in fact, twofold. One line of response was to construct a conceptual account of the resistances that he had run into. He reasoned that he had failed because the time required for mechanical expansion of the chamber was greater than the lifetime of tracks within it; triggering, then, had to fail. This accommodation thus took the form of a mangling—an additive one, in this instance—of his *knowledge* about bubble chambers: he had learned something about them in his practice. And this knowledge hung together with his second line of accommodation, which was to revise his goal. He abandoned

5. Galison states that, after the failure of conventional triggering, "[m]any attempts then followed," including adding carbon dioxide to the chamber to try to slow the speed of bubble formation (324).

the attempt to use the bubble chamber in cosmic-ray physics and decided instead to put it to work in the accelerator laboratory. There, bunches of particles arrived at precisely timed intervals, so that one could expand the chamber by the clock, and the problem of triggering would not arise.

I will discuss Glaser's work in accelerator physics in a moment, but first I want to emphasize that his departure from cosmic-ray physics served to bring out a social dimension of his practice. Glaser later put it this way: "There was a psychological side to this. I knew that large accelerators were going to be built and they were going to make gobs of strange particles. But I didn't want to join an army of people working at the big machines . . . I decided that if I were clever enough I could invent something that could extract the information from cosmic rays and you could work in a nice peaceful environment rather than in the factory environment of big machines . . . I wanted to save cosmic-ray physics" (323–24). The peace/factory opposition in this quotation points to the tension between two distinct forms of work organization that can be discerned in the physics of the early 1950s, "small science" and "big science." Small science was the traditional work style of experimental physics—an individualistic form of practice, requiring only a low level of funding obtainable from local sources and little in the way of collaboration, and promising quick returns on personal initiatives. Big science was the new work style and organizational form that had been born in the US weapons laboratories of World War II and that made its presence strongly felt in the early 1950s in accelerator laboratories like E. O. Lawrence's at Berkeley, Luis Alvarez's base. Big science was done by teams of physicists and engineers, hierarchically organized; it was characterized by a high level of funding and the bureaucratic processes associated with that, a high degree of interdependence in obtaining access to accelerator beams and in the conduct of experimental research, and a relative lack of flexibility in its response to individual initiatives. As the quotation makes clear, Glaser's attachment was to small science, an attachment, as we shall see, that he maintained even as he moved into accelerator-based physics, the home of big science.[6]

With the triggering problem sidestepped, Glaser still faced one difficulty. His prototypes were very small devices, around 1 inch in linear

6. For detailed histories of the emergence of big science in particle physics, see Heilbron and Seidel 1990; and Hermann et al. 1987, 1990. On the history of big science more generally, see the contributions to Thackray 1992 and Galison and Hevly 1992.

dimensions. They demonstrated the possibility of detecting particle tracks, but in themselves they could not compete in data-production rates with other kinds of detectors already in use at accelerators. The key variable was again their mass as a target: Glaser's prototypes simply failed to put enough stuff in the path of the beam. The question was, then, how to scale up the bubble chamber. This was partly a question of its linear dimensions, but also a question of the working fluid: the denser the fluid the smaller a chamber of given mass could be. Glaser had initially, for convenience, used ether as his working substance. At Berkeley, Alvarez opted to work with liquid hydrogen since data taken on hydrogen were the most easily interpreted, and the low density of liquid hydrogen implied the construction of a relatively large chamber (eventually 72 inches long). This was the route that led directly into big science, the route that Glaser wanted to avoid, which he did by seeking to construct a liquid xenon–filled chamber. Since xenon was much denser than hydrogen, he reasoned that a considerably smaller chamber than Alvarez's could be constructed that would still produce interesting physics. His goal with the xenon chamber was to find "one last 'unique niche that I could [fill] at Michigan without access to all this high technology and large engineering staffs'" (327).[7]

Work proceeded on the xenon chamber, but when it was first tested, "no tracks were obtained" (Brown, Glaser, and Perl 1956, 586). Once more a resistance had interposed itself between Glaser and his goal, a resistance that in fact persisted. The paper continues: "The chamber was then altered in an attempt to produce tracks in xenon by enlarging the expansion channel to expand the chamber faster, and by replacing the original Teflon gaskets and diaphragm by butyl rubber to avoid the ex-

7. The xenon chamber was not Glaser's only project in this period. A 15-cm chamber filled with propane that he built and used at the Brookhaven Cosmotron accelerator is recognized as the first bubble chamber to be used in accelerator experiments (rather than as a demonstration device; see Glaser 1964, 539; and Slätis 1959, 13). The attraction of propane chambers was that they operated close to room temperature (60°C) and were therefore much simpler to construct and use than liquid-hydrogen chambers (which operated close to absolute zero), while they contained a greater density of hydrogen atoms than liquid hydrogen itself (Bugg 1959, 19). Propane thus promised to keep Glaser out of the clutches of big science in much the same way as xenon. The drawback with propane was that it did not constitute a pure hydrogen target (due to the presence of carbon nuclei in propane's chemical constitution), which meant that data taken on propane required considerable interpretation to bring them into relation with the concerns of elementary-particle physics.

cessive absorption of xenon exhibited by the Teflon"—the material contours of the chamber were thus mangled, but—"[s]till no tracks were to be seen." This time, though, Glaser and his collaborators found a way around resistance. At the suggestion of colleagues at the Los Alamos laboratory, they tried adding a "quenching" agent, ethylene, to their chamber, and this accommodation was successful. Tracks appeared, and experimentation could get under way (figs. 2.3–2.5).[8] The success of adding ethylene additionally invited a reappraisal of the mechanism of bubble formation by charged particles, and the interpretive account that Glaser had worked with all along was abandoned in favor of the "heat spike" theory formalized by Frederick Seitz in 1958. This last sequence of resistance and accommodation in accelerator physics, then, mangled both the material and conceptual aspects of the culture of particle physics: a new material form of the chamber, the quenched xenon chamber, and new knowledge, a new understanding of the chamber's functioning, emerged together. And further, as I now want to describe, the social dimension of Glaser's practice was mangled in the xenon project, too.

Glaser's research before the switch to accelerator physics had been typical small science. From June 1950 to November 1952—the period that saw the invention and early development of the bubble chamber—he worked in collaboration with a single graduate student, David Rahm, and was supported by the University of Michigan with a total of $2,000. At the end of 1952, the university increased its support to $3,000 per year. This is to be contrasted with the funding and manpower of the xenon-chamber project, where "[p]art-time salaries for Glaser, Martin Perl (a new faculty member), a secretary, and four research assistants added to the salaries for a full-time postdoc and a full-time machinist-technician came to about $25,000. Equipment, supplies, and machining ran about the same" (327). While seeking to propagate the small-science work style, then, Glaser had clearly, in scaling up the chamber and the switch to xenon, evolved something of a hybrid, by no means as solitary and independent as the classic form. But Glaser's xenon project should

8. Glaser's first liquid-xenon chamber had a diameter of 1 inch, was 0.5 inch thick, and was tested in 1956 at the Cosmotron accelerator at Brookhaven National Laboratory (Brown, Glaser, and Perl 1956; Glaser 1958, 327, 340). A 21-liter, 30-cm-diameter xenon chamber was then constructed and was ready for use at the higher-energy Bevatron accelerator in Berkeley in 1958 (Slätis 1959, 16; Glaser 1964).

Fig. 2.3. The body of the 30-centimeter liquid-xenon bubble chamber. Reprinted, by permission, from Glaser 1964, 545, fig. 9. © The Nobel Foundation 1961.

Fig. 2.4. The 30-centimeter liquid-xenon bubble chamber, assembled with ancillary equipment, ready for use at the Berkeley Bevatron accelerator. Reprinted, by permission, from Glaser 1964, 545, fig. 10. © The Nobel Foundation 1961.

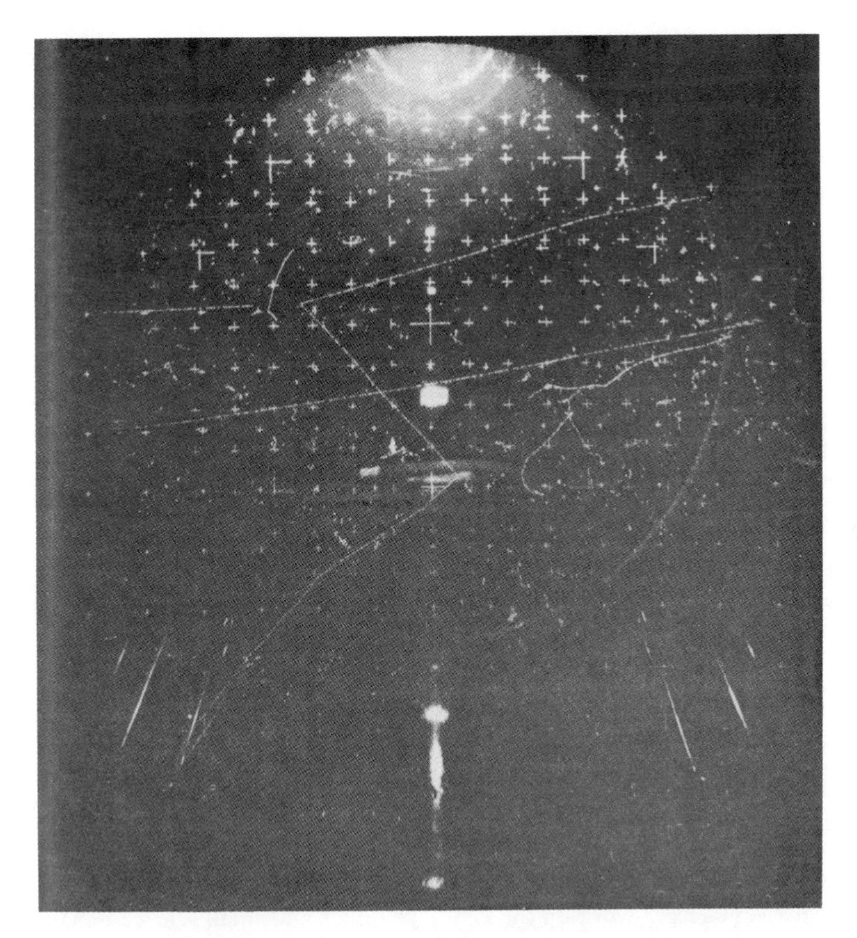

Fig. 2.5. Particle tracks in the xenon chamber. Reprinted, by permission, from Glaser 1964, 548, fig. 13. © The Nobel Foundation 1961.

in turn be compared with Alvarez's liquid-hydrogen effort leading up to the construction of the giant 72-inch chamber at Berkeley (figs. 2.6–2.8). Although the precise extent of Alvarez's empire has never, to my knowledge, been precisely mapped, Galison's account of the work at Berkeley mentions eleven collaborators—physicists, engineers, and graduate students—and it is probably enough to note that (1) the project was eventually funded by the Atomic Energy Commission at $2.5 million, and that (2) Alvarez delegated to Don Gow "[a] new role that is not common in physics laboratories, but is well known in military organizations; he became my 'chief of staff.' In this position, he coordi-

Fig. 2.6. Bodies of liquid-hydrogen bubble chambers of increasing dimensions constructed at Berkeley. The 72-inch chamber is on the right of the front row. Luis Alvarez is second from the right in the back row. Reprinted by permission of the Lawrence Berkeley Laboratory, University of California.

nated the efforts of the physicists and engineers; he had full responsibility for the careful spending of our precious 2.5 million dollars, and he undertook to become an expert second to none in all the technical phases of the operation, from low temperature thermodynamics to safety engineering" (Alvarez 1987b, 125, quoted in part in Galison 1985, 334).[9] In comparison with Alvarez's program, then, which set the standard for bubble-chamber physics in the 1960s, the continued links between Glaser's work on the xenon chamber and small science remain evident.

Before I turn to a general discussion of this passage of practice, two last items of historical information can be included, both of which bear upon the social dimensions of particle physics. Galison notes that "[i]n 1960 Glaser moved to Berkeley to join the growing team of hydrogen-bubble-chamber workers. Shortly afterward, in large part because of his disaffection with the large team, he left physics for molecular biology,"

9. In early 1964 (when the construction phase of the 72-inch bubble chamber was complete and it was in use for experiments), Swatez (1966, 112) counted 23 Ph.D. physicists in Alvarez's group, 20 graduate student research assistants, and about 170 technical and administrative support personnel. I thank Peter Trower for discussions of the work of Luis Alvarez and his group, and for providing the photographs that appear here as figures 2.6–2.8.

Fig. 2.7. The 72-inch bubble chamber ready for action. The upper deck supports the control instrumentation and the liquid-hydrogen supply, the lower deck the chamber, encased in a coffinlike structure. Reprinted by permission of the Lawrence Berkeley Laboratory, University of California.

and "[i]ndeed, by February 1967 Alvarez too had begun to devote almost all his time to other projects, principally his balloon work on cosmic rays" (353). There is a wonderful circularity here, with Alvarez regaining (something like) small science in the field that Glaser had failed to save with the bubble chamber.[10]

10. For more on Alvarez's move into small science, see Alvarez 1987a.

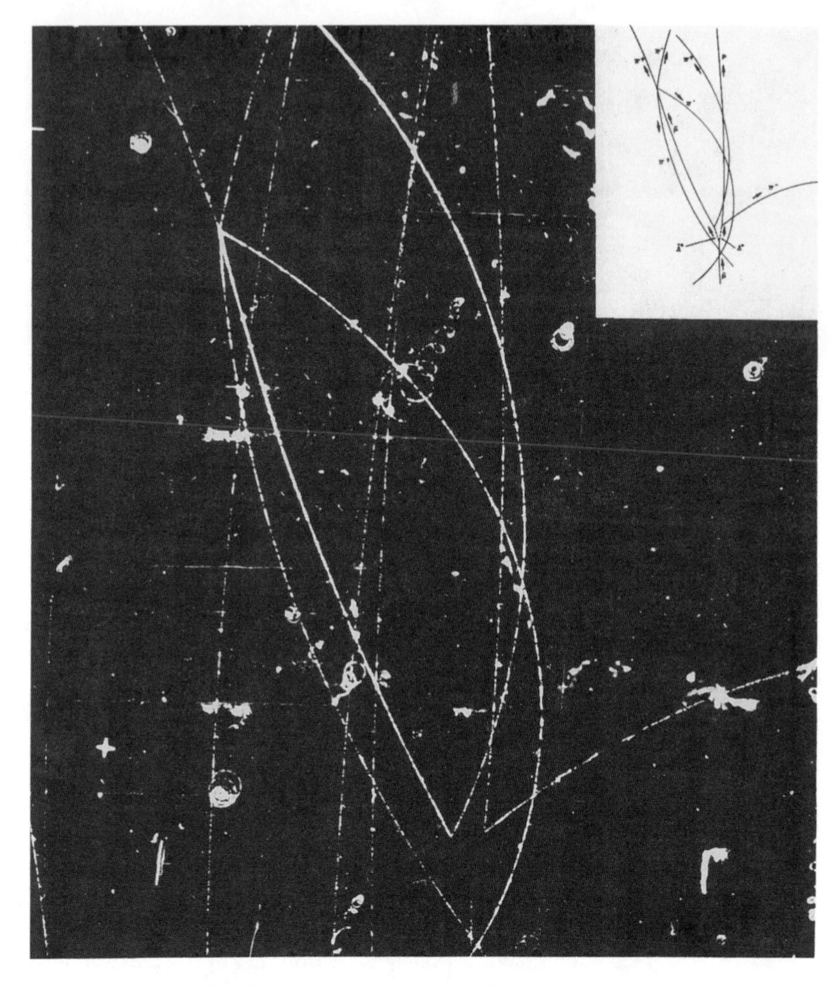

Fig. 2.8. Photograph of tracks in the 72-inch chamber. Reprinted by permission of the Lawrence Berkeley Laboratory, University of California.

2.2 THE MANGLE AND MATERIAL AGENCY

I have told the story of the bubble chamber as an exemplification of the dialectic of resistance and accommodation in scientific practice—the mangle in its most restricted sense—and I have outlined the mangling of the material, the conceptual, and the social in Glaser's practice. Now I need to connect this story to my wider conception of the mangle and my earlier remarks on agency and temporal emergence. I talk first about material agency and then about human agency. Throughout the discussion, the emergent and posthumanist intertwining of material and human agency is evident, but at the end of the chapter I seek to highlight these themes from a different angle, in a comparison of my account of human agency with traditional humanist schemata.

To begin with material agency, we can return to a line of thought introduced in chapter 1. My suggestion is that we should understand the history of the bubble chamber as a more-or-less violent tuning process involving the continual reconfiguration of material setups in the pursuit of an intended capture of material agency. This tuning process was itself organized as a dance of human and material agency. In passages of human activity, Glaser assembled sequences of material apparatuses—the different combinations and configurations of bits and pieces that led up to his prototype bubble chambers, and the various combinations that led from them to the quenched xenon chamber. Interspersed between these bursts of human activity were periods of human passivity, which can be seen symmetrically as periods of material activity. Having assembled some configuration of parts, Glaser stood back to see what they would do.[11] It is worth noting that his tentative assemblages all did something, though typically not what Glaser hoped. His bubble chambers all boiled sooner or later, for example, when the pressure was released, but most of them boiled explosively and uselessly, throughout their volume rather than first along the lines of particle tracks. The dance of agency, then, at once traced out specific machinic configurations *and* their specific powers and characteristics: this combination of parts acted this way, that the other.[12]

Now I can clarify my sense of material agency. It is simply the sense that Glaser's detectors *did* things—boiling explosively or along the lines of tracks or whatever—and that these doings were importantly separate from Glaser. To understand what happened when Glaser took a passive role in the dance of agency, I think one has to acknowledge that some

11. This image of passive watching is quite literal: Glaser stood back with a high-speed movie camera in his hand to record the earliest moments of bubble formation, when tracks might be evident (322).

12. I should add that for practical purposes physicists cared about more than whether their chambers simply produced tracks. They went to considerable lengths to optimize the performance of their chambers, paying great attention to questions concerning the timing of photography, problems of image distortion, and so on. The process of tuning working bubble chambers was thus much more complex than I (or Galison) have described it. For access to the technical details, see, for example, Glaser 1958; Bugg 1959; Slätis 1959; and Weiss 1990.

other source of activity and agency was at work, and in this case that source was the material world (as configured this way or that in Glaser's prior activity). Bubble chambers, when they work, produce tracks and photographs in a way that is not substantively attributable to any human agent. Scientists build and operate chambers, but neither *that* tracks appear in them nor the specific configuration of those tracks is in the hands of the chambers' human companions. Glaser, as far as I can make out, had no choice in the matter: particular material setups performed as they would, and he had to take it or leave it. And this last remark brings us, of course, back to the mangle. From the human point of view, the dance of agency was structured as a dialectic of resistance and accommodation, where resistance amounted to a failure to capture material agency in an intended form, and accommodation to a reconfiguration of the apparatus that might find a way through such resistances. The dialectic of resistance and accommodation—the mangle—thus traced out a trajectory through the space of possible machinic configurations and of their performances, and terminated, in this instance, in successful captures of agency: the particular combination of parts that constituted Glaser's prototype chambers manifested just the kind of material agency that Glaser sought, and later his quenched xenon chamber did the same.

So, my argument is that we should see the bubble chamber as effecting a capture of material agency, as a particular combination of material elements that *acts* in a particular way. The next point that I need to reiterate from chapter 1 is that such captures are themselves temporally emergent. It is clear that Glaser had no way of knowing in advance that most of his attempts to go beyond the cloud chamber would fail but that his prototype bubble chambers would succeed, or that most of his attempts to turn the bubble chamber into a practical experimental device would fail but that the quenched xenon chamber would succeed. In fact, nothing identifiably present when he embarked on these passages of practice determined the future evolution of the material configuration of the chamber and its powers. Glaser had to find out, in the real time of practice, what the contours of material agency might be. This process of finding out is what I have conceptualized in terms of the dialectic of resistance and accommodation, and one can rephrase what has just been said by pointing to the brute emergence of resistance. There is no real-time explanation for the particular pattern of resistances that Glaser encountered in his attempts to go beyond the cloud chamber. In his practice, these resistances appeared as if *by chance*—they *just happened*. It just happened that, when Glaser configured his instrument this way (or this, or this), it did not produce tracks, but when he configured it that way, it did. This is the strong sense of temporal emergence implicit in the mangle.

Here it might be useful to return briefly to Collins and Yearley's dilemma concerning material agency. Their argument was that one has either to think about accounts of material agency as the products of human actors, or about material agency itself along the lines of the scientists' accounts of it (thus ceding analysis of science to the scientists and engineers). As I indicated before, the present analysis of material agency as temporally emergent evades both horns of this dilemma. It recognizes material agency as that with which scientists struggle (unlike SSK), without acceding to scientific accounts of such agency. Having said that, of course, I must pay attention to the fact that Glaser did produce accounts of the functioning of the bubble chamber—knowledge—as he went along. He explained the failure of triggering in terms of the response time of the bubble chamber, and the initial failure of the xenon chamber in terms of a revised understanding of the mechanism of bubble formation.[13] But, as I noted in the introduction, such accounts pose no problem for real-time analysis of practice—they should themselves be seen as part and parcel of the mangling process, as products of the dialectic of resistance and accommodation, at once retrospective glosses on emergent resistances and prospective elements of strategies of accommodation. Their substance can no more be understood in advance of practice than the material contours of nonhuman agency. Both material agency and articulated scientific accounts thereof are temporally emergent in the mangle. Much more, of course, needs to be said about scientific knowledge and the mangle, but I postpone further discussion to chapters 3 and 4, which offer more detailed examples.

One last point concerning material agency. My argument is that we need to recognize that material agency is irreducible to human agency if we are to understand scientific practice. Nevertheless, I need to stress that the trajectory of emergence of material agency is bound up with that of human agency. Material agency does not force itself upon scientists. There is, to put it another way, no such thing as a perfect tuning of machines dictated by material agency as a thing-in-itself; scientists,

13. As an antidote to correspondence realism (section 6.1) about such understandings, it is worth mentioning that the heat-spike model of bubble formation subverted Glaser's prior understanding, which he had relied on in the successful construction of his earlier chambers (328).

to put it yet another way, never grasp the pure essence of material agency. Instead, material agency emerges via an inherently *impure* dynamics that couples the material and human realms.[14] The resistances that are central to the mangle in tracing out the configurations of machines and their powers are always situated within a space of human purposes, goals, plans; the resistances that Glaser encountered in his practice only counted as such because he had some particular end in view. Resistances, in this sense, exist on the boundaries, at the point of intersection, of the realms of human and nonhuman agency. They are irrevocably impure, human/material hybrids, and this quality immediately entangles the emergence of material agency with human agency without, in any sense, reducing the former to the latter.[15] This entanglement is, so to speak, the far side of the posthumanism of the mangle: material agency is sucked into the human realm via the dialectic of resistance and accommodation. Now I turn to the converse proposition—that human agency is itself emergently reconfigured in its engagement with material agency.

2.3 THE MANGLE AND INTENT

> History knows the truth. History was the most inhuman product of humanity. It scooped up the whole of human will and, like the goddess Kali in Calcutta, dripped blood from its mouth as it bit and crunched.
>
> Yukio Mishima, *The Decay of the Angel*

> [W]hat we speak of as a subject's "needs," "interests," and "purposes" are not only always changing, but they are also not altogether independent of or prior to the entities that satisfy or implement them; that is, entities also produce the needs and interests they satisfy and evoke the purposes they

14. Just to be clear, the notion of impurity evoked in this sentence (and later) is relative to my division between human and material agency. The natural sciences are pure, in my sense, in seeking to characterize the pure material dynamics of material agency; the traditional social sciences seek to characterize the pure social dynamics of the social. My argument is that the real-time analysis of scientific practice cannot sustain conceptions of either a pure material or a pure social dynamics. I thank Diederick Raven for encouraging me to clarify this point.

15. One can make a connection to the actor-network here. Where the latter typically speaks of the indistinguishability of people and things, it seems to me more perspicuous to note the liminal position of resistance. Only after resistance has been successfully accommodated does it become possible retrospectively to locate it in, say, the deficient performativity of a machine or a faulty human practice or an unattainable goal or a misarticulated conceptual structure.

implement. Moreover, because our purposes are continuously transformed and redirected by the objects we produce in the very process of implementing them, and because of the very complex interrelations among human needs, technological production, and cultural practices, there is a continuous process of mutual modification between our desires and our universe.

Barbara Herrnstein Smith, *Contingencies of Value*

We can turn from material to human agency. In this section, I focus on the intentional structure of the latter; in the next, on its social structure. In chapter 1, I suggested that in order to understand the work of cultural extension in science one needs to recognize its orientation to specific goals, toward the achievement of specific future states of culture. And I further suggested that such goals should themselves be understood as constructed out of existing culture in a process of modelling. Now I want to connect these remarks with the example at hand.

One can divide Glaser's work on the bubble chamber into two phases. In the first, he sought to create some new type of detector with crudely specified properties based upon the cloud chamber. This phase came to an end with the construction of his small, prototype bubble chambers. In the second, he sought to construct a practically useful chamber based on those prototypes. The cloud chamber and the prototype bubble chambers, then, served as the *models* for Glaser's practice in the two phases of his work: initially he aimed at a detector like the cloud chamber but transformed in variously specified respects; later he aimed similarly at some transformed image of his early bubble chambers.[16] There is nothing special about Glaser's work in this respect. Modelling is, I think, constitutive of scientific practice—in the sense that it is impossible to imagine Glaser embarking on the path that led to the bubble chamber without the example of the cloud chamber before him, or on the path to the quenched xenon chamber without passing through his prototypes.[17] And a recognition of this point implies certain important parallels between and intertwinings of the intentional structure of human agency and material agency. I sketched these out in the abstract in

16. One can continue the list of models. For example, the established practice of triggering cloud chambers in cosmic-ray physics served as a model for Glaser's development of his prototypes.

17. As Krieger puts it, "Analogy is destiny" (1992, 23). "But," he continues, "it is how we analogize that that gives us our destiny back to ourselves," and "just how an analogy is destiny and a metaphor is authoritative are matters that we must figure out for ourselves and check out and discover in the world" (28). These qualifications point, on my analysis, to the mangle.

chapter 1, and now I can return to them as they apply in the present instance.

First, modelling situates human goals and purposes with respect to the cultural field in which they are constructed—the field of existing detectors in the present example. And such cultural situatedness immediately implies a degree of both temporal emergence and posthumanist intertwining in human intentionality. Concerning the former, I argued in section 2.2 that, for example, the precise material configuration and properties of Glaser's prototype chambers were temporally emergent, and the present discussion connects that emergent form to the goals of Glaser's subsequent practice. The goals of scientific practice must therefore be at least as emergent as the models on which they are based. The posthumanist aspect of those goals follows equally directly. Though Glaser formulated the goals of his practice as a classically human agent, the field of existing detectors in which he formulated those goals was a field of material agency. There is, then, a temporal and posthumanist interplay here between the emergence of material agency and the construction of human goals.

We can take this line of thought further in relation to the observation made in chapter 1 that modelling is an *open-ended* process with no determinate destination: a given model does not prescribe the form of its own extension. This openness is nicely exemplified in the history of the bubble chamber: Glaser tentatively imagined and sought to construct a whole range of different kinds of detectors, all modelled, in one way or another, on the basic form of the cloud chamber, before eventually succeeding with the bubble chamber. Likewise Glaser, Alvarez, and the Chicago group sought to develop Glaser's prototypes along quite different axes, and Glaser himself, as we have seen, developed the model in many different ways—first the several versions of the triggered chamber for cosmic-ray physics, and then the variants of the xenon chamber. Modelling, then, is the link between existing culture and the future states that are the goals of scientific practice, but the link is not a causal or mechanical one: the choice of any particular model opens up an indefinite space of modelling vectors, of different goals. And the question therefore arises of why particular scientists fix upon particular goals within this space. Here, as already indicated, I have no principled suggestions. One can speak of scientific creativity, or one can say that the formulation of goals just happens: it just happened, for example, that Glaser set himself the goal of going beyond the cloud chamber, that along the way he hit upon the idea that led him to the bubble chamber, and so on. Certainly nothing identifiably present in advance determined the intentional structure of his practice. Again we run into a role for chance and brute temporal emergence in scientific practice, and we need, therefore, to think about the scientists' goals as being emergent in at least two senses: practical goals are constructed in a temporally emergent cultural field, and their detailed substance is itself emergently constructed in that field.[18] But more remains to be said. We need to think about what goal-oriented human practice looks like in its temporal extension.

Most of Glaser's practice did not consist in the formulation of goals; it consisted rather in material attempts to achieve such goals. He knew, more or less vaguely, the kind of detectors that he wanted to construct, and he spent most of his time trying to make them. The latter consideration, of course, returns us to the mangle, to the dialectic of resistance and accommodation in the engagement with material agency. And the point that I now need to emphasize is that accommodation can amount to a revision of plans and goals, to a revision of the intentional structure of human agency. When Glaser gave up the attempt to trigger his chamber, for example, at the same time he gave up the idea of inserting his new detector into cosmic-ray physics and relocated his goal in the accelerator laboratory. Goals, then, have themselves to be seen as subject to mangling in practice.[19] This observation, in turn, brings us to the third and last sense in which I want to describe intentionality as temporally

18. One can say more about goal formation. Elsewhere (Pickering 1981c, 1984b) I have argued that expertise is a key variable to consider in the dynamics of scientific practice: one can think of expertise as among the resources that scientists can deploy in pursuit of ends, and hence as structuring the particular ends that given scientists choose to pursue. Thus Glaser's early expertise in the use of cloud chambers in cosmic-ray physics clearly structured his practice in developing the bubble chamber, as did Alvarez's expertise in big science (see below). I will remark on more instances of such structuring as the book goes on, but two points about expertise need to be recognized: it is itself open-ended, being deployable in an indefinite range of future projects, and it is itself emergent—expertise comes with practice (and in a posthumanist fashion if that practice engages with material agency). The appeal to expertise does not, therefore, yield a determinate account of goal formation. In Pickering 1990b, I approached the problem from another angle, trying to analyze the goals that Alvarez formulated around the bubble chamber in terms of the intersection of modelling vectors and the piling up of cultural resources. Again, I think that this illuminates the processes of goal formation and elaboration, but it does not efface the elements of contingency and temporal emergence present in it.

19. If one thinks through strategies of accommodation in detail, some ambiguity between means and ends becomes evident in scientific practice. Thus it seems reasonable to see Glaser's move from cosmic-ray to accelerator-based physics as a shift in goal, while his moves through the space of possible triggering arrangements seem better described as the

emergent: the transformation of goals in practice has to be understood in terms of contingently formulated accommodations to temporally emergent resistance.[20] And it points, yet again, to the posthumanism of the mangle. As I remarked when discussing material agency, resistance emerges at the intersection of human and material agency, and as the present argument suggests, serves to transform the former in one and the same process as it delineates the latter. Just as the mangle, then, pulls material agency onto the terrain of human agency, so it materially structures the goals of human agency. Just as the evolution of material agency lacks its own pure dynamics, so too does the evolution of human intentions. "Desire only exists when assembled or machined," as Deleuze and Parnet put it—and vice versa.

2.4 THE MANGLE AND THE SOCIAL

In the last two sections, I have been concentrating on the technical extension of scientific culture in relation to my conceptions of material and human agency and their temporally emergent intertwining. So far, though, I have only talked about the intentional structure of human agency, while in chapter 1 I also discussed the mutual tuning and interactive stabilization of human and material performances. There I took human performativity to be exemplified in the disciplined practices that envelop machines, but I postpone further discussion of discipline to section 3.5 (largely because I have little information on relevant historical developments around the bubble chamber). In this section, I focus rather upon the explicitly social dimensions of human performance in science. Specifically, I want to talk about transformations in the scale and social relations of human agency. One appealing line of thought here concerns the circulation of the bubble chamber and its products through the particle-physics community. It would be very interesting to analyze the

---

(fruitless) exploration of various possible means to an already decided (but eventually abandoned) end. This ambiguity does not undermine the present argument: both means and ends are bound up in human intentionality. Suchman's work (1987) on "plans and situated actions" is relevant and informative here.

20. As in the case of goals, I can offer no principled analysis of why given scientists arrive at particular strategies of accommodation. Indeed, on my analysis, goal formation and accommodation have an equivalent status. Thus Glaser's initial goal of constructing some new kind of detector can be seen as a rather drastic accommodation to the problems of collecting data on strange particles mentioned at the beginning of section 2.1.

new social relations among experimenters, and between them and theorists, that were achieved as bubble chambers and bubble-chamber photographs became increasingly central to the practice of elementary-particle physics, and to explore the further transformations in machines, conceptual structures, and disciplined practices that were constitutive of this centrality. But that would take us too far away from the empirical base of this chapter.[21] Instead I will discuss transformations of the social in and around the work of Glaser and, later, Alvarez. (I will also return to the mangling of the social in chapter 5, in the context of a quite different example.)

I have so far discussed Glaser's practice in terms of technical models, the cloud chamber, and the early versions of the bubble chamber. But a different kind of model was also in play, a model of the social organization of research. In the development of the bubble chamber, Glaser clearly modelled his practice on the small-science exemplars with which he was familiar from his earlier work in cosmic-ray physics. He aimed to

21. I can offer a few brief remarks. The notion of circulation points immediately to an aspect of the multiplicity of scientific culture that gets less attention than it deserves in this book; the idea is that, say, bubble-chamber photographs made transits between quite different cultures sustained in different social groups (experimenters and theorists, to pick an extreme example) even while remaining within the overall field of particle physics (for other detailed examples, see Fujimura 1992; and Star and Griesemer 1989). Further, it is clear that Glaser and Alvarez both aimed at such circulation in their practice—it was partly constitutive of their goals, and partly definitive of what could count as a satisfactory capture of material agency. In this sense, as well as that discussed below in the text, they each aimed at a techno*social,* rather than a purely technical, transformation of existing culture. That the contours of the explicitly social aspects of such extensions cannot be foreseen is evident in the fate of Alvarez's chamber. As Galison (337–40) makes clear, Alvarez sought to optimize the dimensions of his chamber for the production and circulation of data on electrically neutral strange particles (the hardest ones to detect). Ironically, while the chamber did prove useful for this purpose, its historical significance lay rather in its sheer size, which proved invaluable in discovering and exploring the properties of the highly unstable "resonance" particles, identified in increasing numbers from the late 1950s onward, that were central to the quark model of the strong interactions (Pickering 1984b). One can make a connection here to the "opportunism in context" model of the dynamics of scientific practice laid out in Pickering 1984b: to see actors as aiming at an imaginative transformation of technosocial culture is one way to grasp the immediately social correlates of individual practice. Pickering 1984b lacks, however, the mangle as the dialectic of resistance and accommodation that lies between scientists and their goals (and it also lacks the concept of material—and disciplinary—agency). I thank Yves Gingras and Dominique Pestre for encouraging me to clarify this point. One can also make a direct connection to Knorr-Cetina's idea (1981) of goals as marking a "promise of success," and the actor-network's idea that scientists construct "scenarios" or "scripts" that they then seek to persuade nonhuman and human actors to act out.

propagate small science into the work of constructing and using bubble chambers.[22] And in this connection, it is important to note that modelling in social practice, as in technical practice, has to be understood as open-ended. Glaser, that is, had to *find out* in practice what "small science" might amount to in the bubble-chamber project. Interestingly enough, as we have seen, this process of finding out transformed Glaser from an almost classic microactor (though even in his early work at Michigan Glaser was assisted by a graduate student) into something of a macroactor, the leader of a team of no less than nine people. I did not discuss this transformation in detail in my narrative, but enough has been said, I think, to tie it into my general analysis of the emergent and posthumanist quality of human agency. Concerning the former, we need only note that no one could have foreseen that this social transformation would come about. No identifiable feature of Glaser's initial situation determined it. Glaser certainly did not intend it at all. Instead, Glaser's small-science model, as I said, was itself open-endedly mangled in practice.[23] And again we can note the posthumanist decentering of this mangling. As discussed in chapter 1, a kind of interactive tuning of the social and the material was in play here; the social evolution of Glaser's work style was itself constitutively the product of maneuvers in the field of material agency. Most strikingly, it seems clear that Glaser's practice would have remained much more individualistic if he had succeeded in triggering his chamber on cosmic rays; likewise, the nine-person team would have collapsed if the xenon-chamber project had failed. To pick

22. In section 2.1, I followed Galison in focusing on the construction phase of Glaser's chambers, but in the development of large instruments, there is an important and interesting break between the phases of construction and use, involving a wholesale reorganization of both technical work and social organization. I have no information on the latter phase of Glaser's work, but Galison (1985) covers the transition in Alvarez's group, and Swatez (1966, chap. 5) offers interesting insights into the day-to-day working of Alvarez's group around the bubble chamber as a finished instrument.

23. Another way to put this point is to note that, while Glaser's interest in small science was clear enough, he was not limited in his practice by any closed definition of it. In effect, he had to find out what would count as small science in the course of his project—or, equivalently, to find out what he was willing to tolerate as close enough to his basic conception of small science. In Pickering 1990b, I make a similar point concerning Alvarez's finding out just what big science could amount to in bubble-chamber physics. As noted at the end of the historical narrative, both physicists eventually decided that the social organization of bubble-chamber work had become intolerable and left the field, but that this was an emergent upshot of practice is especially clear in the case of Alvarez, who had deliberately set out to construct the big-science form of life that eventually repulsed him (Alvarez 1987a).

up an important concept that I introduced in chapter 1, the failure of triggering destabilized Glaser's small-science approach to experimental physics, while the performativity of the quenched xenon chamber and the nine-person collaboration *interactively stabilized* one another: they emerged together, sustaining and sustained by each other, in the real time of the practice of Glaser's growing team.[24] Here as before, then, one needs to think of the impure dialectic of resistance and accommodation between human and material agency in order to comprehend the mangling of the social contours and embodiment of the former. The social aspects of Glaser's practice did not evolve in accordance with any pure social dynamics; no purely sociological explanation can suffice to explain Glaser's transformation toward the status of a macroactor.[25]

My argument is, in effect, that the social dimensions of scientific culture should be seen as themselves in the plane of practice and as always, in principle, subject to mangling there, just like and together with the material and conceptual dimensions. This point is, I think, important enough to merit further development and exemplification, which can be accomplished by shifting briefly from Glaser's to Alvarez's work on the bubble chamber.[26]

If Glaser's model for the social organization of science was a small-science one, Alvarez's was big science. In the early 1950s, he had enormous experience in big-science projects carried out in the wartime and postwar US weapons laboratories, and he saw in the bubble chamber a means of recycling that expertise into experimental particle physics.[27] Part of the attraction of building a large bubble chamber filled with liquid hydrogen was that it would link the kind of large-scale engineering and physics effort that he was familiar with to an instrument that could outgun (to pick an appropriate metaphor) other instruments currently

24. As indicated in chapter 1, my notion of interactive stabilization is tied to a vision of cultural multiplicity, but I postpone further development of this theme to chapter 3.

25. On purity and impurity, see note 14 above. The emergent "coproduction" of social structure and material agency is a central theme in the actor-network approach: for some exemplifications, see Callon 1987; and Latour 1983, 1987.

26. Besides Galison 1985, there is a considerable secondary literature on Alvarez and the bubble chamber: see Alvarez 1987a, 1987b; Heilbron, Seidel, and Wheaton 1981; Pickering 1990b; Swatez 1966, 1970; and Trower 1987, 1989.

27. In World War II, Alvarez worked first at the MIT Rad Lab and then at Los Alamos. After the war, he first became involved in the construction of a large proton accelerator at Berkeley and then in the Materials Testing Accelerator project at what became the Livermore National Laboratory. The latter was another military project, which had just collapsed when Alvarez became interested in bubble-chamber work.

in use. As in the case of Glaser, though, the point that I want to dwell upon here is that Alvarez, too, had to find out in practice what big science might amount to in the context of his project.[28] Most interestingly, from the perspective of the present discussion, on their path to the 72-inch chamber, Alvarez and his collaborators found it necessary to their program to build novel social and technical links to groups outside the experimental particle-physics community. Thus, for example, quite early in the project, problems became apparent both in producing and working with large quantities of liquid hydrogen. These problems were addressed in the establishment of a collaboration with a group from an Atomic Energy Commission–National Bureau of Standards laboratory in Boulder, Colorado, that was already in possession of the appropriate hardware and expertise (334–37). Likewise, as the program continued and the specifications of the 72-inch chamber were drawn up, it became clear that data-production rates for the chamber were going to be greatly in excess of existing methods of data analysis. This resistance—this data bottleneck as Galison calls it (340)—was accommodated by the incorporation of computers, computer programmers, scanning equipment, and nonphysicist scanners into the business of experimental physics in a quite new way. Again, the important point to bear in mind here is that these specific features of big-science experiment as it evolved at Berkeley were not intended from the start. They emerged as part of a real-time dialectic of resistance and accommodation, and the new relations established between the Berkeley team, the AEC/NBS laboratory in Boulder, and the worlds of computing and paid labor need to be seen as interactively stabilized in relation to the emerging contours and powers of the 72-inch chamber.[29]

This is as far as I want to take the discussion of human agency for

28. While the business of building and running particle accelerators had already become big science by the early 1950s, experiment around accelerators still retained the traditional small-science form. Thus there were no models for big science in experimental-particle physics that Alvarez could draw directly upon. It was nevertheless the case that experimenters in accelerator laboratories had to deal with their big-science aspects: hence Glaser's aversion to entering this arena.

29. A further degree of symmetry needs to be mentioned here. Just as Alvarez's work helped to define what "big science" would be in elementary-particle experiment, so it helped to define the nature of the other actors with whom he made alliances. Particle-physics research was no part of the Boulder laboratory's mission until it was enrolled by Alvarez (as Callon and Latour would put it)—previously his contacts there had been engaged in the (abandoned) attempt to build a "wet" hydrogen bomb—and the same remark applies to the emerging world of electronic computers, programmers, and so on.

the moment. Its upshot is that both the "inner" social organization of human agency (its scale, as embodied in micro- or macroactors: Glaser) and its "outer" organization, too (the relations between different actors: Alvarez) are in the plane of practice and liable there to emergent and posthumanist mangling, interactively stabilized (and destabilized) alongside and in association with the material (and conceptual) dimensions of scientific culture.[30]

2.5 ACTORS, INTERESTS, AND CONSTRAINTS

To complete this first exemplification, it might be useful to thematize the emergent posthumanism of the mangle from a different angle, so I close with a brief comparison of my account of human agency with traditional humanist accounts. The latter fall into two classes. One class—encompassing, for example, pragmatist and symbolic interactionist approaches—explicitly recognizes the temporal emergence of human agency, and differs from my account only in its human-centeredness.[31] I do not need to discuss it further. The other class, though, is less explicit about emergence, to say the least, and this is the class that I focus on. It can itself be subdivided into two rather different understandings of the intentional structure of human agency. On the one hand, following Marx and Weber, human goals and purposes can be characterized positively in terms of, say, the *interests* of individuals

30. Examples of modelling in the extension of the (techno)social culture of science can be multiplied indefinitely from the literature on the history of science, though they are often not thematized as such, and the level of detail required to recognize the mangling of the social is often not present. As just one example, and to stay close to the present material, it is interesting to trace out the ways in which the Second World War weapons laboratories in the US were modelled on E. O. Lawrence's prewar Radiation Laboratory at Berkeley and how the weapons laboratories in turn constituted the models for postwar reorganization of academic physics: see Heilbron and Seidel 1990; Kevles 1987; Galison 1988b; and Schweber 1992a. There is a fascinating literature in the symbolic interactionist/pragmatist tradition of science studies on the creation and maintenance of links between different "social worlds" within and beyond science: see, for example, Fujimura 1992; Star and Griesemer 1989; and Star 1991a, the last being good on the asymmetric manglings that the production of such links can involve. A concern with changes of scale in actors (including the emergence of new ones) and with their reconfiguration in the establishment of relations with one another is definitive of the actor-network approach to STS (though I want to reconceive this concern in a performative rather than a semiotic idiom). See Callon and Latour 1981; and Latour 1983, 1987.

31. On pragmatist and symbolic interactionist understandings of human agency, in science and more generally, see the works cited in chapter 1, note 33.

and groups. On the other, following Durkheim, they can be characterized negatively, in terms of *constraints* on human action.

Both of these perspectives on human agency have been articulated within SSK: David Bloor (1983, 1992), for example, develops a basically Durkheimian stance; Barry Barnes (1977, 1982), Steven Shapin (1979, 1982, 1988a), and Donald MacKenzie (1981a, 1981b) favor versions of the interest model. I will discuss specific examples in sections 4.6 and 5.3 and the general explanatory strategy of SSK in section 6.6, but for the moment I can note more generally that my account of human agency has resonances with both interest and constraint models. Most obviously, my insistence that scientific practice has to be understood as goal-oriented tends to align my analysis with the interest model. Some care is needed here, though. While it is not, I think, a necessary feature of the interest model that interests be understood as nonemergent causes of practice, it remains the case that the canonical SSK literature on the interest model has, in fact, shown little interest in how interests themselves change in practice. The tendency is to write as if the substantive interests of actors were present and identifiable in advance of particular passages of practice, setting them in motion and structuring outcomes without being themselves at stake.[32] I therefore need to make it explicit that, if my analysis of practice is to be understood as a posthumanist variant of the interest model, that can only be on condition that the goals of science are understood as subject to impure mangling, as already described. They cannot be regarded as unmoved movers, as causal principles lying outside (behind, above) and explaining the extension of scientific culture.[33] Further, one is accustomed to thinking of interests as attached to actors, micro or macro, and, again, the canonical SSK litera-

32. Thus in stating the first tenet of the strong program in SSK, Bloor (1992, 7) invites us to look for the social causes of belief, but in none of the four tenets does he invite us to explore transformations of the social.

33. To make their point about "social construction," most studies in SSK understandably focus on instances where interests arguably remain constant through practice. This is the kind of situation in which traditional nonemergent sociological analyses can be seen to do explanatory work, but my point is that it is not exemplary of the general situation. In contrast, I take Latour's discussion (1987, 108–21) of the "translation" and "enrollment" of interests to point in the same direction as the present discussion, though his concern with the attempts of human actors to work specifically on the interests of others is not integral to my account—the changing intentional structure of Glaser's practice, for example, cannot be understood in terms of his direct interactions with other human beings. In critical theory, Smith (1988) spells out an emergent and posthumanist understanding of the intentional structure of human action (see the quotation at the beginning of section 2.3), likewise Deleuze and Guattari (1987) in philosophy and Graham (1994) in

ture has shown little interest in the transformation of actors in scientific practice, thus tending to enforce by default an image of their identities as nonemergent and persisting unchanged through practice. In contrast, I hope that the discussion of the preceding section has made it clear that the social identities and relations of scientific actors—their inner and outer constitution—are themselves liable to mangling and redefinition in practice.[34]

To speak of the mangling of actors and their goals is, of course, to invoke the dialectic of resistance and accommodation, and my appeal to resistance as an explanatory category clearly puts me somewhere near the terrain of the second accounting scheme mentioned above, the one that understands human agency in terms of constraint. Importantly, though, a clarification of the difference between my conception of resistance and traditional notions of constraint can serve to foreground what is novel about the mangle.[35]

The point is this. While the word "constraint" can be given many meanings, it seems to me to be endowed with two quite specific features in the humanist schema. First, it is located *within* the distinctively human realm. It consists, say, in a set of social (or epistemic) norms, derived in some sense from social structure. And second, it is discussed as temporally *nonemergent,* at least on the time scale of human practice. Constraints are continuously present in culture, even when not actively operative. The language of constraint is the language of the prison: constraints are always there, just like the walls of the prison, even though we only bump into them occasionally (and can learn not to bump into them at all).[36] My usage of resistance has neither of these qualities. As I have emphasized, in the real-time analysis of practice, one has to see

social theory. For a fascinating discussion of "interest" and related concepts in early German social theory, see Turner 1991.

34. See also Latour's discussion of the construction of new social groups with new properties (1987, 115–16). The classic work in social history on the construction of actors and their interests is E. P. Thompson's study (1963) of the making of the English working class. Sewell (1992) discusses the problems created by the concept of "structure" in social theory for the analysis of social change, and offers a constructive critique of the work of Anthony Giddens and Pierre Bourdieu, which moves some way in the direction suggested here. See also Knorr-Cetina 1988.

35. I thank Michael Lynch and others who, in professing to see no difference between resistance and constraint, forced me to think this issue through. I discuss specific examples of constraint and limit talk in sections 5.3 and 6.5.

36. Thus Giddens conceptualizes the whole Durkheimian tradition like this: "The structural properties of social systems . . . are like the walls of a room from which an individual cannot escape but inside which he or she is able to move around at whim"

resistance as genuinely emergent in time, as a block arising in practice to this or that passage of goal-oriented practice. Thus, though resistance and constraint have an evident conceptual affinity, they are perpendicular to one another in time: constraint is *synchronic,* preexisting practice and enduring through it, while resistance is *diachronic,* constitutively indexed by time.[37] Furthermore, while constraint resides in a distinctively human realm, resistance, as I have stressed, exists (at least as we have discussed it so far) only in the crosscutting of the realms of human

(1984, 174). In contrast to this picture, Giddens's structuration theory "is based on the proposition that structure is always both enabling and constraining" (169). Hence, the previous quotation continues, "[s]tructuration replaces this view [of complete freedom within a room] with one which holds that structure is implicated in that very 'freedom of action' which is treated as a residual and unexplicated category in the various forms of 'structural sociology.'" The walls are still there, though. (Giddens later abandons metaphor for something close to tautology: "What, then, of structural constraint? . . . it is best described as *placing limits upon the range of options open to an actor*" [176–77]). For a poetic variation on the theme, see Ginzburg 1980, xxi: "In the eyes of his fellows, Menocchio was a man somewhat different from others. But this distinctiveness had very definite limits. As with language, culture offers to the individual a horizon of latent possibilities—a flexible and invisible cage in which he can exercise his own conditional liberty." An invisible rubber prison with very definite limits. As Shapin (1988a, 549 n. 14) notes, even Latour is not immune from this style of nonemergent thinking; as, for instance, in his idea that "[i]nterests are elastic, but like rubber, there is a point where they break or spring back" (Latour 1987, 112–13). From my perspective, Wise (1993, 249) muddies the waters by introducing the concept of an "elastic resistance"—"the network acquires an elastic resistance to deformation palpable to anyone who would enter it from outside or who would threaten it"—but immediately drops it in favor of "constraint." I discuss Galison's "contexts and constraints" image of scientific practice in section 6.5.

37. A conversation with Michael Power helped me to see the mangle as a temporally rotated version of traditional accounts of human agency. I hope that pointing out that resistance performs a similar role in my analysis to constraint in traditional accounts makes it clear that my emphasis on temporal emergence does not amount to "anything goes." Comments from, among others, Irving Elichirigoity, Peter Galison, and Paul Forman have encouraged me to make this point explicit, and I return to it in the discussion of objectivity in section 6. 4. I should further make it explicit that I have no objection to the notion of constraint as an actors' category. Certainly, actors often do construe their situations and develop their practice in terms of articulated notions of constraint. I suggest, however, that such accounts need to be analyzed as constructed in practice and themselves subject to mangling—just like any other item of knowledge (as in Pickering 1981a). Constraints should not be "ontologized"—they should not be treated, as is often done, as somehow structuring and thus explaining the flow of practice from without. Constraints are as emergent as anything else. Thus, to give one example, in his early work, Glaser found that the interior of bubble chambers had to be extremely clean if tracks were to be formed. This became an element of bubble-chamber lore and can readily be understood as a constraint on chamber development. Interestingly, though, this lore came under pressure, especially at Berkeley, where it was evident that the big chambers that Alvarez had in mind would necessarily be "dirty" ones (in a technical sense, meaning having

and material agency. Resistance (and accommodation) is at the heart of the struggle between the human and material realms in which each is interactively restructured with respect to the other—in which, as in our example, material agency, scientific knowledge, and human agency in its intentional structure and its social contours, are all reconfigured at once. Coupled with the rotation in time just mentioned, this displacement—from constraint as a characteristic of human agency to resistances on the boundary of human and material agency—serves to foreground the emergent posthumanist decentering implicit in the mangle.[38]

---

metal-to-glass joints). At this point, one of the Berkeley technicians, A. J. Schwemin, just ignored the constraint and went ahead, building a relatively large dirty chamber, which proved to work quite satisfactorily. This constraint was discontinuously mangled—it disappeared—in material practice (Alvarez 1987a, 118–19).

38. Note that one can arrive at notions of emergence within traditional humanist accounts of human agency via the introduction of feedback. Interests and constraints might be seen as subject to conscious or unconscious revision in response to practical experience. This line of thought has not been well developed in science studies, but within sociology more generally it is familiar in, for example, the guise of Anthony Giddens's structuration theory (1984), where the idea is that social structure both constrains (and enables) action and is itself subject to reflexive monitoring and modification in the practice of individual agents. Law 1993 and Law and Bijker 1992 borrow from structuration theory in the sociology of science and technology; the latter essay construes the entire range of studies collected in Bijker and Law 1992 as emergent. For a valuable history of feedback thought in the social sciences, see Richardson 1991. I thank Bruce Lambert for encouraging me to think about structuration theory in relation to studies of scientific practice. Of course, no amount of feedback can transform the asymmetric attribution of agency that is definitive of traditional humanist approaches.

THREE

Facts

The Hunting of the Quark

My text, my word, my body, the collective with its agreements and struggles, bodies that fall, flow, burn or resound just as I do, all these are only a network of primordial elements in communication with each other . . . nature is formed by linkings.

Michel Serres, *Hermes*

In the last chapter, I presented my first empirical example of scientific practice, and I laid out my basic understanding of the mangle as the emergent intertwining of human and material agency in a dialectic of resistance and accommodation. In line with my interest in rebalancing our understanding of science toward its material and performative dimensions, the example concerned the development of an instrument rather than the production of scientific knowledge per se. I did show that the early history of the bubble chamber encompassed the production of a certain kind of knowledge, and that the contents of that knowledge could be understood as products of the mangle. Nevertheless, the knowledge in question had literally as well as metaphorically an instrumental character. It was knowledge of the inner workings of the bubble chamber (concerning its response time and the mechanism of bubble formation) rather than of the inner workings of nature. The latter, however, is what we are accustomed to regard as exemplary of scientific knowledge, and in this chapter I want to explore directly how scientific knowledge, now explicitly referring to nature rather than to machines, can be integrated into my machinic picture and performative idiom. Especially I am interested in the production of scientific facts—empirical statements about the world—and their relation to scientific theory.

I proceed again by example, analyzing the practice of the Italian physicist Giacomo Morpurgo in a program of experiments that he conducted at the University of Genoa between 1965 and 1981, experiments that sought to find evidence for or against the existence of peculiar elementary particles called quarks. While important to the history of particle physics, these experiments were simple and straightforward, both conceptually and materially, compared with most experiments at the research front of the natural sciences, which is what recommends them for examination here.[1]

In section 3.1, I review the history of the Genoa experiments. I note that the production of empirical knowledge here entailed the open-ended extension of both material and conceptual elements of scientific culture, and that Morpurgo's practice was organized around the construction of *associations* or alignments between them that would lead outward from his material apparatus and its performance into the world of articulated knowledge and representations. My central point is that such alignments are not easily made, and that their successful construction happens, if at all, as the upshot of a dialectic of resistance and accommodation, of the mangling of both material and conceptual culture. The alignments at issue, then, serve at the same time to sustain particular items of scientific knowledge and to single out and interactively stabilize particular vectors of cultural extension in material and conceptual space. At the material end, I note that beyond the capture of material agency such stabilizations involve a precise *framing* of machinic performances in relation to the conceptual structures with which they are aligned.

1. Beyond their technical simplicity, another advantage of studying these experiments is that Morpurgo's writings, published as well as unpublished, display a sensitivity to the temporal dimension of practice that is rare among scientists. I have been interested in Morpurgo's work since the late 1970s, when I undertook a study of the controversy between him and William Fairbank, to be discussed in section 6.6 (Pickering 1981b), and I want to thank Professor Morpurgo for his assistance over the years, in correspondence, in the provision of published and unpublished material, and, most recently (1992), in making it possible for me to visit his laboratory and to see what remains of his apparatus. I should make it clear, however, that Professor Morpurgo offers no endorsement of what follows. He wrote of my earliest account of his work that "[w]e certainly appreciate the intention of the study of Dr. Pickering; however we disagree with many of his statements" (Marinelli and Morpurgo 1982, 258), and this continues to be the case for the present chapter. My understanding is that his unease has two connected sources. From the physicist's point of view, the account that follows is historically unbalanced. In concentrating on early phases of Morpurgo's program, it tends to emphasize episodes that appear trivial in retrospect at the expense of the final, difficult, and highly sophisticated ferromagnetic experiment which I discuss relatively briefly. More generally, Professor Morpurgo has asked me to state that he "does not share many aspects of my general view of the interrelationship between theory and experiment" (letter of 22 November 1993).

In section 3.2, I summarize what we have learned and emphasize the emergent and posthuman quality of Morpurgo's practice and its products. In section 3.3, I generalize the discussion, by observing that Morpurgo's practice exemplifies an idea that I mentioned in section 1.1 but have so far left undeveloped, namely, that scientific culture is *disunified, multiple, and heterogeneous.* Thus Morpurgo worked at once upon both conceptual and material elements of his culture. Further, as just mentioned, he worked to create associations between these heterogeneous elements, to bring them into relation with one another, in the production of facts. My suggestion in section 3.3 is that the production of such associations in a multiple and heterogeneous culture is a general feature of scientific practice, by no means confined to this example, and the next two sections develop this suggestion in two directions. In section 3.4, I draw upon additional studies to argue that in the production of facts and the relation of facts to theory much more multiplicity in the conceptual stratum of science is typically at stake than is evident in the case of Morpurgo's quark searches. I argue that we should therefore think of scientific knowledge as sustained in extended *representational chains* spanning multiple levels of theoretical abstraction, and that alignments along such chains should themselves be understood as subject to, and the products of, mangling in practice. In section 3.5, I turn from multiplicity to heterogeneity, and discuss another cultural stratum—that of disciplined human practices—that is in play in the production of empirical knowledge (and everything else) in science. Once more invoking additional studies, my argument is that disciplined practices, too, should be understood as mangled and interactively stabilized together with the material and conceptual strata of scientific culture. The upshot of this chapter is thus that scientific knowledge should be understood as sustained by, and as part of, interactive stabilizations situated in a multiple and heterogeneous space of machines, instruments, conceptual structures, disciplined practices, social actors and their relations, and so forth. This is my version of Serres's idea that "nature is formed by linkings."

One last introductory comment, concerning the social dimensions of scientific practice. In what follows, I tend to speak of Morpurgo in the singular, but this proper name—like "Glaser" in much of chapter 2—should be taken as the designation of a small macroactor of changing composition, comprising Morpurgo himself and a varying collection of students and junior colleagues. One can discern significant manglings of the composition of this group as the quark-search experimental program evolved, but none was as striking as those around the bubble chamber, and for clarity I confine to footnotes my discussion of the collaborative aspects of Morpurgo's work.[2]

3.1 THE HUNTING OF THE QUARK

Before we get to the Genoa experiments, some technical and historical background is needed, starting with the physicists' conception of a quark (see Pickering 1984b for the details). The story begins in the early 1960s, the time of the "population explosion" in the world of elementary particles. Before then, the existence of only a few distinct particles had been recognized, but suddenly the list of known particles—strictly speaking, of strongly interacting particles, or "hadrons"—started to increase fast (in large part due to data gathered using Luis Alvarez's 72-inch bubble chamber, discussed in chapter 2). In 1961, theorists Yuval Ne'eman and Murray Gell-Mann constructed an ordering scheme known as SU_3 or "the eightfold way," which promised to introduce some order into this proliferation of particles by sorting them into families or "groups." In 1964, Gell-Mann and George Zweig took the SU_3 idea one step farther by suggesting that the basic SU_3 group structure could itself be explained if hadrons were understood as, in some sense, composed of combinations of more fundamental entities that Gell-Mann called quarks. Quarks, then, might be the true elementary particles, the fundamental building blocks from which the cosmos was built. In the present context, however, the most important thing to know about quarks is that they were supposed to carry *fractional electric charges*—third-integral ones, either $e/3$ or $2e/3$, to be precise. This was in contrast to the received belief in physics since the beginning of this century that all electric charge is quantized in units of e, the charge on the electron. Everything from elementary particles to macroscopic chunks of matter had hitherto been confidently supposed to carry electric charges that were some integral multiple of e (or zero), a supposition backed up by years of experiment.

2. My information on the structure and division of labor within Morpurgo's group is taken from his letter to me of 29 April 1993. In the following notes, I indicate the particular kinds of expertise that different physicists brought to the Genoa experiments as exemplification of the "opportunism in context" model of scientific practice developed in Pickering 1984b.

Gell-Mann's and Zweig's 1964 proposal accommodated this apparent conflict between the properties attributed to quarks and the findings of experiment by suggesting that quarks were always to be found bound together to form hadrons in just the right combinations such that the third-integral charges of the constituent quarks added up to an integral multiple of *e*. There was also a possibility, however, that occasionally one might come upon a quark in isolation, since a straightforward argument led to the conclusion that the lightest quark (there were supposed to be three distinct species) would be stable—there would be no way in which it could get rid of its fractional charge.[3] And an experimental agenda could be readily constructed from this observation. Measurements of electric charges—either on elementary particles or on macroscopic samples of matter—would bear rather directly upon the quark hypothesis: any evidence for third-integral charge would immediately be evidence for the existence of "free" or "isolated" quarks. During 1964, several physicists reasoned along these lines, and some of them actually embarked on experimental searches for free quarks. Among this band was Giacomo Morpurgo, who in the end pursued free quarks more intensely, over a longer period, than anyone else. In what follows, I want to explore the development of his experimental program and its findings over time.

In 1965, Morpurgo and two colleagues constructed an instrument that they came to call a magnetic levitation electrometer (MLE), and we will be concerned with this instrument and how Morpurgo used it to construct facts. Before we come to that, though, I can insert one remark. Morpurgo's path to the MLE was not simple or straightforward. Like Donald Glaser, Morpurgo began with a crude vision of the sort of detector he needed to construct, but this vision was radically mangled in his early practice. I could indeed tell the story of the genesis of the MLE along just the same lines as I told the story of the genesis of the quenched xenon bubble chamber in the last chapter—in terms of the formulation of specific goals and plans in processes of modelling, dialectics of resistance and accommodation organized around intended captures of mate-

3. The first person to mention this possibility was Gell-Mann, though he did not take it seriously. He concluded his original quark publication by remarking that "[a] search for stable quarks . . . would help to reassure us of the non-existence of real quarks" (1964, 215). Gell-Mann and his followers tended to regard quarks as convenient mathematical constructs.

rial agency, which radically transformed both Morpurgo's plans and the material form of his apparatus, and so on. But, just because this part of the story reproduces the features that I analyzed in chapter 2, I will not go through it here.[4] I want to concentrate on the fact-producing phase of Morpurgo's experiments, which means, for the sake of clarity, starting the story with the MLE already in place.

Morpurgo's experiment was designed to look for fractional electric charges on macroscopic samples of matter in a bench-top laboratory experiment. He aimed to do something like the classic Millikan oil-drop experiment, in which the charge of tiny drops of oil is measured by observing their responses to an applied electric field. In the 1910s, Robert Millikan had done much to establish the quantization of electric charge using this approach (Holton 1978), and Millikan's experimental procedure had since become part of the staple diet of teaching laboratories. Since free quarks had not been experimentally observed in such setups, however, they had to be very rare in the terrestrial environment—if they existed at all—and Morpurgo therefore aimed somehow to scale up the procedure in order to investigate much larger quantities of matter than the tiny droplets of oil traditionally examined. Millikan's experiment, then, was Morpurgo's basic model, though it had, in fact, been transformed almost beyond recognition on the way to the MLE.

The MLE was built by Morpurgo and two collaborators, who submitted their first report for publication on 23 July 1965 (Becchi, Gallinaro, and Morpurgo 1965).[5] It was designed to measure electric charges carried by small grains of graphite (soot). These were levitated in vac-

4. The key text on Morpurgo's route to the MLE is Morpurgo 1972, an essay written for the series *Adventures in Experimental Physics* (which ceased publication before the essay could appear) and explicitly intended to display the vicissitudes of experimental practice rather than to edit them out, as is customary in scientific writing (see Pickering 1989b). I discuss the initial phases of Morpurgo's practice briefly in Pickering 1989c, 283–85; and Gooding (1990, 193–203; 1992, 80–91) constructs graphic maps from an earlier unpublished account of mine based on Morpurgo 1972. Working through the prehistory of the MLE in an earlier draft of this chapter first convinced me of the necessity to include material agency in my understanding of scientific practice.

5. Prior to the quark-search experiments, Morpurgo's entire professional career had been in theoretical physics. He was one of the first theorists to take the quark model seriously. Unlike Gell-Mann and his followers, Morpurgo adopted a "very naive" and literal way of thinking about quarks, conceptualizing them as heavy, slowly moving, physical constituents of hadrons, and found that several surprisingly accurate predictions of empirical quantities could be obtained in this way (Morpurgo 1965; Becchi and Morpurgo 1965a, 1965b). This was what encouraged him to think seriously about experimental

uum in a magnetic field: graphite is diamagnetic, meaning that it is repelled from regions of high magnetic field, and using an electromagnet with appropriately shaped pole caps, Morpurgo found that it was possible to float the grains in a magnetic "well" (fig. 3.1). Achieving such a stable suspension system—a capture of material agency in my terms—was the hardest part of getting the first version of Morpurgo's experiment running, the locus of many manglings. Bracketing the position at which the grains floated were two small metal plates. Voltages could be applied to these plates, and the response of the grains to the electric fields thus created could be observed with a microscope. The magnetic suspension system, the metal plates and their power supply, plus an optical system and various ancillary components, constituted the material form of Morpurgo's apparatus (figs. 3.2–3.5).

To move from observations of the response of the grains to an applied electric field to statements about the electric charges carried by the grains, Morpurgo drew upon his *interpretive account* of the MLE, his conceptual understanding of how it functioned.[6] The basic form of this was very simple, being given by the laws of classical electrostatics. According to the latter, the force F on a charge q due to an electric field E is $F = Eq$. E could be calculated directly as the ratio between the measured voltage V, applied to the plates divided by the distance between them, d. Morpurgo's basic interpretive account of the MLE was thus the equation $F = (V/d) \times q$. Through this equation, a measurement of F could be translated directly into a measurement of q, the charge on the graphite grain being observed. And F itself could be measured by observing the extent of the lateral displacement of the grain away from its equilibrium position at the bottom of the magnetic well (P_0 in fig. 3.1) when

searches for free quarks (Morpurgo 1972, 1–2; in Pickering 1984b, 117–18 n. 32, I explore the relation between Morpurgo's work on the quark model and his prior work in nuclear and elementary-particle physics). C. Becchi was a young theorist who collaborated with Morpurgo in some of the early theoretical work on quarks and who joined him in the quark-search experiment. In 1967, Becchi was involved in military service, after which he declined to work further on the quark experiment and worked only in theoretical physics. It was Becchi who introduced Morpurgo to Gaetano Gallinaro, a young experimenter—indeed the only experimental physicist in the original collaboration. Morpurgo recalls that Gallinaro's expertise in vacuum technology was particularly important in the first magnetic-levitation experiments.

6. In my earliest account of Morpurgo's experiments along the lines developed in this chapter (Pickering 1989c), I spoke of "interpretive models" in this connection; I use the phrase "interpretive account" here to avoid possible confusion with my notion that practice is constitutively a business of modelling.

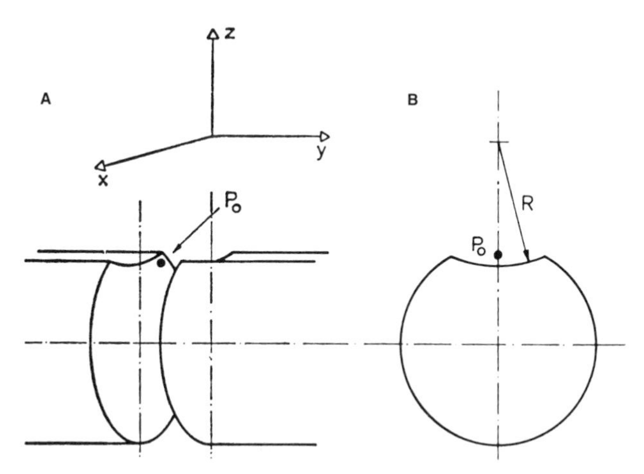

Fig. 3.1. Magnetic levitation: (A) perspective drawing of the pole caps of Morpurgo's magnet; (B) cross-section of the pole caps. Diamagnets are repelled from regions of high magnetic field, and the departures from circularity at the top of the poles create a magnetic "well" in which graphite particles can be suspended. P_0 is the equilibrium position of the particles. Reprinted, by permission of Elsevier Science and G. Morpurgo, from Morpurgo, Gallinaro, and Palmieri 1970, 96, fig. 1.

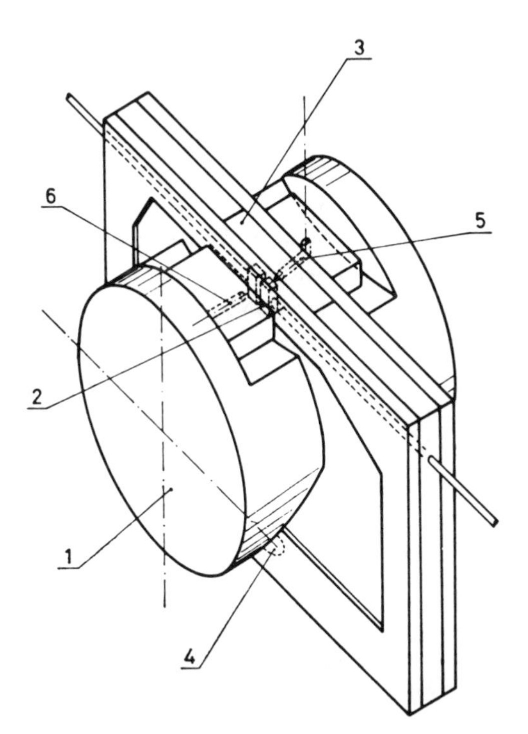

Fig. 3.2. A schematic drawing of the core of the magnetic levitation electrometer (around 6 inches overall in its linear dimensions). Key elements are the pole caps of the electromagnet (1) and the tiny plates used to apply the electric field (2). The grains were observed from above through a window (3). Reprinted, by permission of Elsevier Science and G. Morpurgo, from Morpurgo, Gallinaro, and Palmieri 1970, 109, fig. 10.

the electric field was turned on. Morpurgo reasoned along routine lines that the distance the grain was displaced in space was linearly proportional to the force acting upon it.[7]

So much for the material form of Morpurgo's instrument and his interpretive account of it. Now we can turn to the history of its use. Here is Morpurgo's own description of the earliest measurements.[8]

> I recall clearly the measurements on the first grain. We had spent the whole morning trying to find an appropriate grain with a small value of the charge and at the end we had been lucky. We had a grain which, on applying the electric field, moved by seven divisions of our graduated scale. We had lunch and after, maybe, two hours we came back. The grain was of course still there. Only on applying the same electric field it did not move anymore; or better it did move by only one division, and the *same* movement, in the *same* direction it had when the electric field was reversed. (This meant, of course that there was a gradient in the electric field, since the force due to such a gradient EδE/δx is unchanged if the field is reversed.) Clearly the grain in these conditions was neutral; during the lunch time it had captured an ion from the residual gas and the charge had probably passed from one to zero. Because I had to go to a faculty meeting we postponed the beginning of the measurements to the end of the meeting; when, after three hours, I came back the grain was in the same charge state as before; and in about one hour we could perform the measurements which are reported in the first column of the table I of [our first publication, Gallinaro and Morpurgo 1966].
>
> I have indulged in this description because we were very proud of the fact that, aside from having already reached a sensitivity a few hundred times larger than Millikan, our method was much more convenient in that there was no need of observing continuously the grain.
>
> In the next few days we performed the other five measurements reported in the table; there was no indication of fractionary charges, so far,

7. I suppress some detail here. In order to calibrate the displacements of the grains from their equilibrium positions, Morpurgo varied their charges by exposing them to a weak radioactive source, designed to change their charges by small multiples of the charge on the electron. The "step length" corresponding to unit change in charge could then be used to calculate the "residual charge" on each grain—the charge that came closest to zero. A residual charge of zero corresponded to an integrally charged grain; a residual charge of ± 1/3 corresponded to an isolated quark on the grain.

8. I quote from Morpurgo's unpublished first-person account (see note 4 above) because it is more vivid and contains more temporal detail than the published account, Gallinaro and Morpurgo 1966.

Fig. 3.3. The magnetic levitation electrometer. The circular components in the center of the picture are the magnet poles; the microscope looks down from above onto the viewing window (held in place by eight bolts). Reprinted, by permission of Elsevier Science and G. Morpurgo, from Morpurgo, Gallinaro, and Palmieri 1970, 107, fig. 8b.

and we had already performed the equivalent of one thousand Millikan measurements (the standard Millikan droplet weighs 10^{-11} g.). We therefore decided to write a letter to Physics Letters. (1972, 5–6)

With this passage, culminating in the publication of the empirical fact that no fractional charges were to be found on the graphite grains so far inspected (Gallinaro and Morpurgo 1966), we can begin our examina-

Fig. 3.4. The entire magnetic levitation electrometer, including ancillary equipment. The large circular components at the bottom are the coils of the electromagnet. In this version, the microscope has been connected to a TV camera (at the top of the picture) for remote viewing. Reprinted, by permission of Elsevier Science and G. Morpurgo, from Morpurgo, Gallinaro, and Palmieri 1970, 106, fig. 8a.

tion of how Morpurgo threaded his material apparatus into the field of articulated scientific knowledge and vice versa. Even from this brief and simple description of his opening moves, one can see that facts did not flow mechanically in a predetermined manner from the MLE. Much work remained to be done, work that I want to argue amounted to a

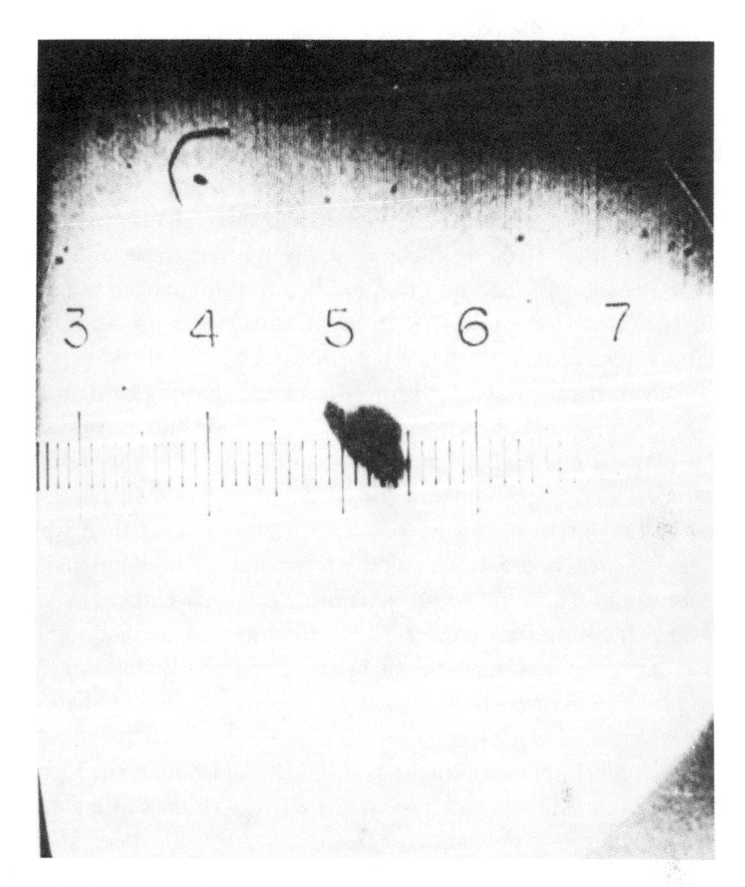

Fig. 3.5. A picture of a levitating graphite grain as relayed to a TV set. The microscope scale used to measure displacements of the grain from their equilibrium positions is clearly visible. Reprinted, by permission of Elsevier Science and G. Morpurgo, from Morpurgo, Gallinaro, and Palmieri 1970, 117, fig. 14.

fascinating interplay and emergent mangling of the material and conceptual dimensions of Morpurgo's practice.

The first point to note is that the practice described in the above passage has once more the character of a dance of human and material agency. As a classic human agent, Morpurgo assembled his apparatus, switched it on, and then, surrendering his active role, stood back to watch what would happen—literally, through a microscope. Swapping roles, the material world was in turn free to perform as it would: the grains levitated and moved away from their equilibrium positions when the electric field was applied. And immediately a problem arose. The very first grain acted strangely. This grain, Morpurgo recalled, displayed "the *same* movement, in the *same* direction" when the direction of the applied electric field was reversed, and the point to note is that this behavior was quite anomalous when understood in terms of Morpurgo's interpretive account of his instrument as formulated above. According to that, the force on a given charge was supposed to change its direction when the direction of the field changed. Straightforwardly understood, then, the behavior of the first grain implied that it changed the sign of its electric charge—from positive to negative or vice versa—but not the magnitude of its charge whenever the direction of the field was reversed. Such behavior made no sense within Morpurgo's existing field of understandings of electrical phenomena, and counted for him as a *resistance* on the path to his goal of charge measurement. And this resistance prompted a further exchange in the dance of agency. Morpurgo once more took the initiative, but this time in conceptual rather than material practice. His accommodation took the form of a revision and mangling of his interpretive account of his apparatus. In his parenthetical remark beginning "[t]his meant, of course," he explains and justifies a revision to his interpretive account—the inclusion of an additional term in the formula for calculating charges to allow for electric field gradients acting on the grain.

Various points are worth making about this episode. First, I have already noted that Morpurgo's basic interpretive account was a simple—in fact the simplest—application of classical electrostatic theory. Schoolchildren learn that the force on a charge is Eq. What becomes immediately apparent in the history of Morpurgo's experiment, however, is that the development of an interpretive account of a real piece of apparatus, even one as materially simple as Morpurgo's, is an open-ended process. As we will see, Morpurgo open-endedly extended his basic interpretive account in a variety of ways as his program of experiments evolved; the addition of a term to allow for electric field gradients was just the first such move. The question arises, therefore, of how we should understand Morpurgo's trajectory through the indefinite space of possible electrostatic interpretive accounts. Clearly he did not explore this space at random. Not just any mathematical formula could count—for him or his colleagues—as a plausible interpretive account. One has, I believe, to think about the special dynamics of conceptual practice in this connection, about how Morpurgo sought to tie the particular interpretive accounts that he fixed upon into the general frame of established knowledge in electrostatics and prior applications thereof. I want to postpone discussion of this topic until I can give it detailed attention in chapter 4. For now, I want just to note that (1) conceptual practice has to be seen as involving its own nontrivial mangling—this is an important sense in which all of Morpurgo's interpretive accounts were far from arbitrary—but that (2) this conceptual mangling manifestly failed to specify the precise form of interpretive account that Morpurgo should adopt: the sheer variety of models that he defended in the course of his experimental program is sufficient, I think, to establish this.

The problem of understanding how Morpurgo proceeded through the space of possible interpretive accounts thus remains, and one observation is crucial. The amendment to Morpurgo's interpretive account just discussed was *stabilized,* as I say, by the fact that it led to findings that made sense. And to explain more precisely what that means, the introduction of another piece of terminology will be useful. Not just one but two kinds of conceptual account figured constitutively in Morpurgo's practice. Beyond his interpretive account, he also worked in terms of *phenomenal accounts*—meaning conceptualizations of the aspects of the material world that his experiments were designed to explore. In fact, he worked in terms of just two, well-defined and dichotomous, phenomenal accounts: either all of the charges in his laboratory were integral (the received belief) or some of them were third-integral (quarks). And, in effect, Morpurgo aimed at an *association* between the material and conceptual elements of his practice that would amount to one of *translation,* in which his interpretive account would translate the performance of his material instrument into one of his pair of phenomenal accounts.[9] This was how the MLE was to be linked outward into the realm of articulated facts and knowledge. As we have seen, Morpurgo's primitive interpretive account failed in practice to sustain such an association—this is the sense in which the behavior of the first grain constituted a resistance, and in which his accommodating revision to his interpretive account was successful. This success in turn singled out and stabilized (temporarily) this particular extension of Morpurgo's interpretive account from the open-ended field of possibilities. The revised account hung suspended, as it were, between specific material performances on the one hand and the pair of phenomenal accounts on the other.

9. "Translation" is a term of art in the actor-network approach (which sometimes goes under the alternative designation of the "sociology of translation"); my usage is a very literal exemplification of the concept.

We can develop this line of thought by continuing to track Morpurgo's practice. Not content with the charge measurements first published, Morpurgo displaced his goal further into the future, in a pattern repeated throughout his program, seeking to measure charges on ever-increasing quantities of matter—either to improve his chances of finding isolated quarks or to set more stringent upper limits on their abundance in the terrestrial environment. And, in another repeated pattern, this displacement of his goal destabilized the association that he had made between his material procedure and his conceptual accounts. Continuing to make measurements on further samples using the same setup, this is what happened:

> A few days after having submitted it [the report for publication in *Physics Letters*] we found the first "anomalous" event. The grain number 7 looked as if it contained a quark. On reversing the electric field when the grain was in its state of minimum charge it had a displacement, in the same direction, but not equal to the previous one; the difference in these displacements corresponded to one fourth (or with "some goodwill" to one third [i.e. a quark]) of the difference in displacements when the object had captured an electron.
>
> For a few days we were rather excited; the excitement decreased however, when, after a while, we saw that several other grains had a similar, but not qualitatively identical behavior. We finally understood what was going on; the clue came when we decided to make measurements, on the same "anomalous" grain, at different increasing separation of the platelets. In fact the first of these measurements gave the following result: we had a grain showing a residual charge of 1/9 when the platelets had been placed at a separation of 1.6 mm.; this residual charge decreased to less than 1/32 when the separation between the platelets was increased to 2.7 mm. (the applied voltage was also increased to keep the electric field the same). We concluded that we had been observing a spurious charge effect. (Morpurgo 1972, 6–7)

Some more steps in the dance of agency are evident here; the mangle was once more in action in fact production. As Morpurgo surrendered his agency to the MLE, he again encountered puzzling material performances, performances that did not translate through his current interpretive account into either of his phenomenal accounts. Despite having made no changes in his material instrument, grains started to turn up that appeared to have neither integral nor third-integral charges. Morpurgo's observations on these grains seemed rather to point to the conclusion that electric charge was continuously divisible. And this was a result Morpurgo was not prepared to accept. He found it conceivable that the odd, rare quark might exist, but he had not embarked on his program of experiments in the expectation of finding that standard lore on the quantization of charges was entirely misconceived. Within this frame of expectations, therefore, his latest findings counted once more as a resistance: the three cultural elements that he sought to bring together—this particular material instrument, the latest version of his interpretive account, and his pair of phenomenal accounts—interfered with one another instead of hanging together. Once more, Morpurgo took up the reins of agency, and his attempted accommodation to this latest resistance was situated not in conceptual but in material practice. In a continuation of exchanges of activity and passivity between the human and nonhuman worlds, Morpurgo experimented with changes in the separation of the metal plates used in his instrument to apply the electric field ("platelets" as he calls them above), and he found that when this separation was increased the findings of nonintegral charges went away (no quarks, again). The material world thus obliged him with a performance that could be reconciled with his expectations.

A point that I need to stress here is that the engagement of Morpurgo with the material world in this episode differs significantly from those that we examined in chapter 2. Its object was no longer a straightforward capture or downloading of material agency; Morpurgo had already effected this capture when he began taking measurements. Instead it amounted to a further level of detailed tuning of his material instrument, which took the form of a delicate *framing* of material agency. Beyond cajoling the material world into levitating graphite grains for him and moving them around in response to the application of an electric field, Morpurgo had now succeeded in revising the precise material contours and performance of his instrument to fit it into the detailed conceptual schemes provided by his interpretive and phenomenal accounts.[10]

One further point about the framing of material agency. Morpurgo's practice just described had the effect of stabilizing a specific material

10. Bastide (1990, 207–8) uses "framing" (and "focusing") in a similar sense to mine, but in a representational rather than a performative idiom. She is concerned with the artful selection of visual representations to integrate them into textual fields, while I am concerned with the integration of material performances and conceptual accounts.

configuration of his apparatus. The performance of the reconfigured apparatus now translated through the interpretive account into one of Morpurgo's pair of phenomenal accounts. But it is worth emphasizing that the interpretive account that effected this translation was itself precisely the product of a prior transformation and stabilization (in response to the prior performance of the MLE). By this stage of his experiment, then, Morpurgo had reached a point where his material instrument and his interpretive account of that instrument reinforced and stabilized one another: the specific shape and performance of the MLE were underwritten by the fact that the latter translated into the no-quark phenomenal account, and the specific form of the interpretive account was guaranteed by its efficacy in effecting that same translation. Morpurgo's instrument and his interpretive account of it thus *interactively stabilized* one another: each propped up the other in the production of facts. And, of course, the specific fact that Morpurgo reported—the absence of free quarks on a specified quantity of matter—was constituted in this interactive stabilization. The facts came as part of a material-conceptual package.

This point about the interactive stabilization of the material and conceptual elements of science in fact production is the main one I want to make in this section, and I now want to reinforce it with more historical detail. First, I can note that Morpurgo's conceptual analysis of his apparatus went further than I have so far indicated. As it happens, he had reasons for working with the plates close together in the earliest version of the MLE. As he and his colleagues wrote in a definitive report of their experiments carried out up to 1970:[11]

> The question, of course, arises of why we have not always operated from the start keeping the platelets at a very large distance . . . We plan to do this in the future . . . [But] we were led to think to the possibility of the Volta force, while we were working in the static version, keeping the platelets at a small distance. Precisely it was seen that an apparent residual charge effect, present in some cases, did disappear when operating on the same grain at a larger (double) distance. It was felt necessary to explore how general this behavior was . . .

11. G. Palmieri was an associate professor in Morpurgo's department. He was an electronics expert who built several electronic components for the resonance version of the quark-search experiment mentioned below.

> It should be added that our initial tendency was to work at a relatively small distance between the platelets: this tendency was motivated by the argument that the gradient of the applied field should be smaller in this case; it was realized only later that this argument was not only false for small separations between the platelets due to the formation of irregular graphite depositions on the platelets, but also is irrelevant for the Volta force, the main part of which does not depend on the gradient of the applied field. (Morpurgo, Gallinaro, and Palmieri 1970, 103)

At this detailed level of conceptual analysis, therefore, Morpurgo's transformation of the material configuration of his apparatus went against his interpretive account of it, and the success of the former led, in turn, to a further revision of the latter. To explain the success of increasing the plate separation, Morpurgo rejected his earlier calculations, complicating his interpretive account still further to allow for small, irregular, enduring electric fields on the surface of the plates (due to the so-called patch, or Volta, effect, as well as to the piling up of soot on them). This complication enabled him to explain why his earlier findings of all sorts of different charge values were spurious.[12] And here we see yet another variant of Morpurgo's interpretive account being ground out by the mangle, exemplifying further the open-endedness of his representations of his apparatus and the interactive stabilization of particular variants.[13] Amusingly, this open-endedness (and the corresponding lack of necessity of any given interpretive account) can be further em-

12. "Its explanation [the finding of nonintegral charges] is the following: the total force on a grain is $QE + \alpha E\delta E/\delta x$. The first term in this expression is the force on the grain due to its charge (which is what we intend to measure) while the second term is the force due to a gradient in the electric field. On reversing the direction of the applied electric field the two forces behave differently; the first changes its sign, the second does not; this is the basic reason why the experiment can be done also in our geometry where the electric field is far from uniform. However the total field E acting on the grain cannot really be identified with the field E_a we apply. In between the platelets there exists, in addition, a very small electric field E_v due to Volta effects which we cannot reverse (also if the platelets are fabricated of the same metal, they are not monocrystals, and moreover there are always irregular deposits of graphite on them). Let us write $E = E_a + E_v$; then in the previous formula the gradient term produces, besides the main term $E_a\delta E_a/\delta x$ also cross terms of the form $E_a\delta E_v/\delta x + E_v\delta E_a/\delta x$; these cross terms are linear in the applied electric and therefore cannot be distinguished from a charge force on reversing E_a. They simulate a spurious charge; they can be decreased only by increasing the distance between the platelets; indeed this is what we had done" (Morpurgo 1972, 7).

13. Like Glaser's, this mangling of Morpurgo's interpretive account had a prospective as well as a retrospective quality. Morpurgo's calculations indicated that the perturbing

phasized in another quotation from Morpurgo and his colleagues: "Note that it is not clear that the estimates given . . . when extrapolated to distances as small as 2a = 1.6 mm do have much meaning. We repeat in fact what it has already been stated . . . : it is possible, that at these small distances between the platelets, the gradient of the applied field is occasionally larger than the estimate (contrary to what one should expect) due to the irregular depositions of graphite on the plates" (Morpurgo, Gallinaro, and Palmieri 1970, 113).

It is evident enough, therefore, that the key to fact production within Morpurgo's experimental program was not possession of an unarguable conceptual understanding of how the apparatus functioned, as traditional perspectives on science might encourage us to think. It was, instead, the achievement of three-way interactive stabilizations between Morpurgo's material procedures and his conceptual accounts, interpretive and phenomenal, in which maneuvers in the field of material agency played a constitutive role alongside conceptual ones, and material performances and conceptual understandings guaranteed and underwrote one another, back and forth.

So far I have been concentrating on the early phases of the Genoa experiment, up to the end of 1966. As mentioned already, though, experimentation continued for another fifteen years or so, with Morpurgo continually displacing his goal toward the examination of ever larger quantities, and different kinds, of matter. I will not go into detail concerning the later stages of the experiment, but I do want to draw attention to some of the more conspicuous material and conceptual manglings that emerged there. After 1966, the MLE went through two major mutations. In the experiments so far discussed, Morpurgo operated with constant electric fields, observing the constant displacements of his samples from their rest positions when the field was applied. In 1967, he began a new series of experiments using oscillating electric fields to induce the samples to resonate about their rest positions. This enabled him to make accurate observations on much larger samples. Continuing in the same direction, in the 1970s he substituted a ferromagnetic sus-

effects at issue were proportional to the volume of the samples being measured. And thus it was clear that if even larger samples were to be examined the plates would have to be farther apart than was possible in the diamagnetic setup. This was a line of reasoning that led to the ferromagnetic experiments mentioned below, where there was no such limitation on plate separation. (I thank Professor Morpurgo for clarifying this point.)

pension system for his earlier diamagnetic one.[14] While the latter was inherently stable, the former was unstable and required the installation of a sophisticated feedback system to keep the samples steady, but at the same time made it possible to levitate yet heavier samples, now of iron and steel.[15]

At each step along the way, the mangle was in action; the dialectic of resistance and accommodation in material and conceptual practice continued to play itself out.[16] Thus, to give two examples, in the ferro-

14. This amounted to a substitution of exemplary models for the suspension system. The model for diamagnetic suspension had been the pioneering work of Braunbeck (Morpurgo 1972, 3); ferromagnetic suspension had been explored by Jesse Beams, and already put to use in experiments intended to measure the electron-proton charge difference by Stover, Moran, and Trischka (1967).

15. The most detailed account of the resonance version of the diamagnetic experiment is Morpurgo, Gallinaro, and Palmieri 1970. The corresponding account of the ferromagnetic experiment is Marinelli and Morpurgo 1982. Mauro Marinelli initially worked in the industrial firm of Italsider, but in the late 1960s and early 1970s also studied physics at the University of Genoa, taking courses with Professor Morpurgo. According to Morpurgo (letter, 29 April 1993), "he asked me the thesis for his final degree (Laurea); he would have liked a thesis on theory. However at that time I was convinced (and still am) that there were too many Elementary Particle theorists in the world, and I thought that his experience with industry should have been exploited also in his thesis . . . So I assigned to Marinelli a thesis on the levitation with feedback of ferromagnetic substances . . . and . . . he brilliantly succeeded in showing how on big spheres this system could be constructed and worked." Marinelli graduated in 1972 and joined Morpurgo's department as an assistant professor, where he became one of the principals of the most sophisticated version of the quark-search experiment. The other principals were Morpurgo himself and, up until around 1977, Gallinaro (who, Morpurgo recalls, provided important expertise in the use of lock-in amplifiers). On the connection between the switch to ferromagnetic suspension and the earlier mangling of the interpretive account of the diamagnetic experiment, see note 13 above.

16. Some of these manglings are best understood as seeking to optimize the capture rather than the framing of material agency, along the lines already discussed in chapter 2. To give some examples, it turned out that the ferromagnetic levitation electrometer was sensitive enough to be significantly disturbed by traffic noise; measurements were therefore typically carried out overnight, when traffic was less. Illumination with white light proved to induce undesirably frequent charge changes on the samples, to which Marinelli and Morpurgo responded by switching to a red light for the optical system. The overnight readings were taken automatically, and at one time Marinelli and Morpurgo were puzzled by a series of failures, when balls mysteriously fell out of suspension in their absence. Their first thought was that the electrical power supply must be at fault, but no interruptions were apparent when they looked into this. A solution emerged one morning when Marinelli arrived to find a dead fly in the apparatus. He reasoned that flies interrupting the line of sight of the optics had been incapacitating the feedback system that held the samples steady in space, and the successful accommodation to this latest resistance proved to be to hang flypaper around the laboratory (Marinelli, conversation, 4 May 1993).

magnetic experiments of the late 1970s, Morpurgo found that the charges of iron cylinders seemed to drift over time—from zero to *e*/10, for example. Tinkering once more in material practice, Morpurgo found a new way to frame material agency, discovering that he could achieve stable measurements, again of zero charge, if he spun the cylinders. And, tinkering with his interpretive account, he then found that he could make sense of this in terms of "small torques [that] can arise under the action of the oscillating electric field." With interactive stabilization reestablished, Morpurgo reported once more his inability to find evidence for isolated quarks (Gallinaro, Marinelli, and Morpurgo 1977, 1257). Continuing to make measurements, Morpurgo yet again started finding unacceptable results: charges that were all over the place (figs. 3.6 and 3.7). This provoked a drastic revision to his interpretive account. After thinking for a decade and a half about a simple classical electrostatic account of a simple piece of apparatus, he announced the discovery of a new force at work in his apparatus: a magnetoelectric force, coupling the measurement and suspension aspects of his apparatus in a hitherto unexpected way, and capable of mimicking charges up to four times that of the electron. When this force was included in his interpretive account and new material procedures were devised to measure it (another framing), interactive stabilization was restored, and the absence of quarks on a total mass of 3.7 milligrams was reported (Marinelli and Morpurgo 1982, 236; see fig. 3.8). This corresponded to a gain in sensitivity of around ten million over the standard Millikan-type experiment. At this point, around 1981, Morpurgo felt that he had tried hard enough to find isolated quarks and brought his program of experiments to an end.[17] My historical narrative has to end at this point, too, and the rest of the chapter takes this history as its point of departure.

17. Marinelli and Morpurgo (1984) reanalyzed their earlier charge measurements to set an upper limit on the electron-proton charge difference of $10^{-21}e$. Interestingly, from the point of view of the "opportunism in context" model of the dynamics of scientific practice that I developed in Pickering 1984b, the early, diamagnetic-suspension version of the quark-search experiment refused to die. When I visited Professor Morpurgo in May 1993, I was able to see the original equipment still in use, now for levitating grains of the high-temperature superconductors that have been of great interest in physics in recent years. The two principal differences from the quark-search experiment were that (1) a liquid-nitrogen cooling jacket had been added to the apparatus, to bring the samples down to their superconducting temperature, and (2) the veterans of the quark experiment had now been joined by a physical chemist, G. L. Olcese, with expertise in the preparation of high-temperature superconductors. See Marinelli, Morpurgo, and Olcese (1989) for an account of this reincarnation. I am grateful to Professor Olcese for letting me loose on the apparatus while he was trying to work with it.

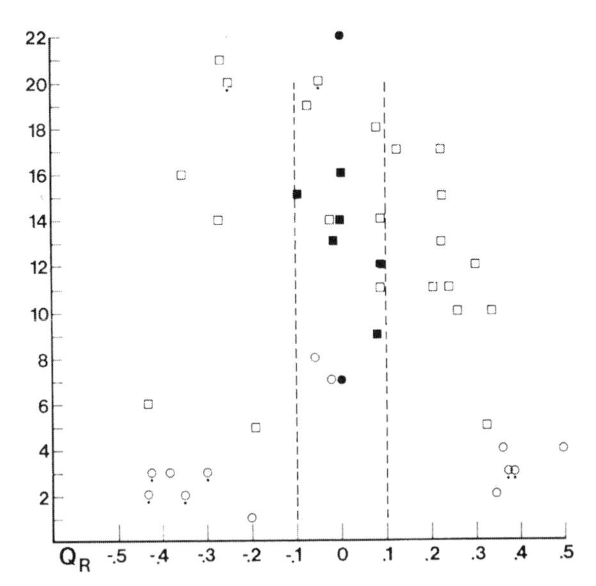

Fig. 3.6. Apparent charge distributions from an intermediate stage of the ferromagnetic experiment. Charges were measured on twenty-two balls. The x-axis gives electric charge values; the y-axis designates particular balls by a number (thus measurements on the same horizontal level are all on the same ball). Reprinted, by permission of Elsevier Science and G. Morpurgo, from Marinelli and Morpurgo 1982, 190, fig. 6.1.

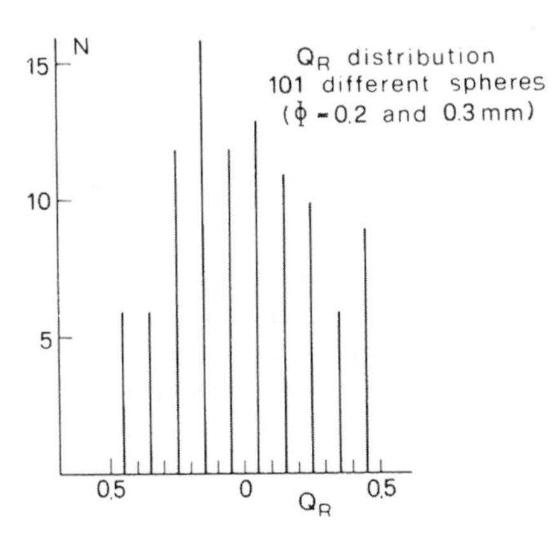

Fig. 3.7. Histogram of charge measurements on 101 different balls. The x-axis measures charge; the y-axis designates how many balls were found to have charges within a specified range. Reprinted, by permission of Elsevier Science and G. Morpurgo, from Marinelli and Morpurgo 1982, 190, fig. 6.2.

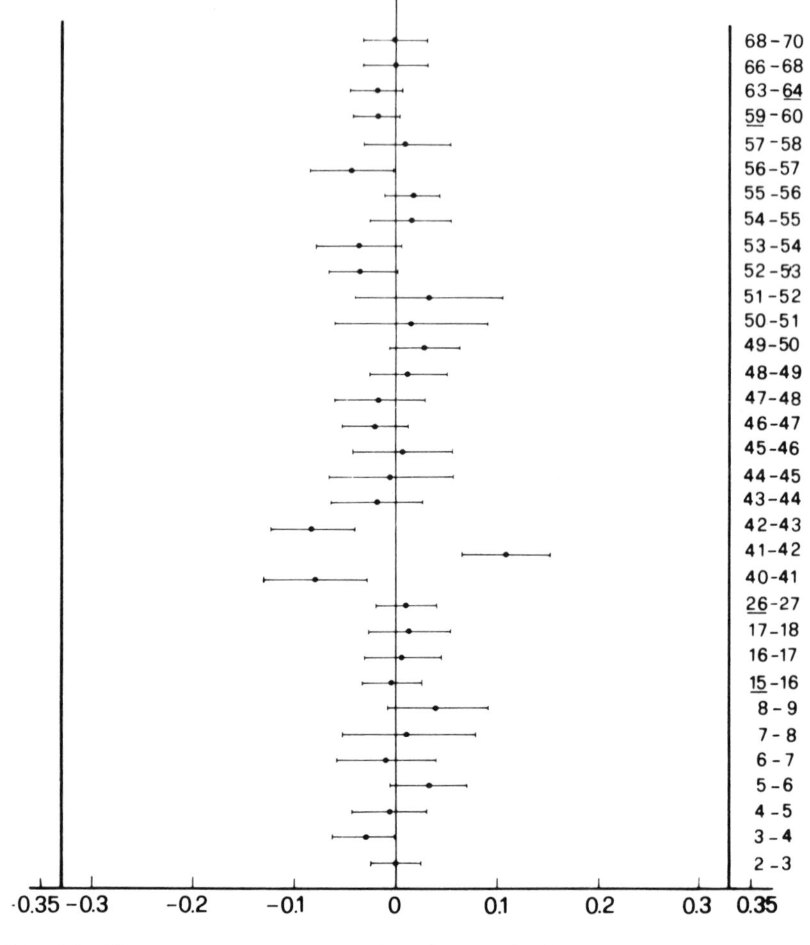

Fig. 3.8. Charge measurements (x-axis) taken in the ferromagnetic experiment after the magnetoelectric force had been recognized and taken into account. Note that the data points lie close to the vertical line corresponding to zero charge and are distant from the lines representing charges ±1/3. Reprinted, by permission of Elsevier Science and G. Morpurgo, from Marinelli and Morpurgo 1982, 233, fig. 8.10.

3.2 EMERGENCE AND POSTHUMANISM IN EMPIRICAL PRACTICE

[N]o one, not even God, can say in advance whether two borderlines will string together or form a fiber, whether a given multiplicity will or will not cross over into another given multiplicity, or even if given heterogeneous elements will enter symbiosis, will form a consistent, or cofunctioning, multiplicity susceptible to transformation. No one can say where the line of flight will pass.

Gilles Deleuze and Félix Guattari, *A Thousand Plateaus*

I can summarize the preceding narrative by saying that in his quark-search experiments Morpurgo open-endedly extended the material and conceptual cultural elements of his practice in a series of stabs—trying first *this* material configuration of his instrument, then *that* interpretive account, and so on, with no advance guarantee of success. The typical upshot of this process was incoherence; the performance of the MLE did not translate through Morpurgo's current interpretive account into one of his phenomenal accounts; resistance was evident relative to the association that would connect happenings in the Genoa laboratory into the world of facts. Occasionally, however, accommodations proved successful. The latest version of the interpretive account would translate the current performance of the MLE into one of the phenomenal accounts. And the significance of such success was twofold. On the one hand, it served to interactively stabilize specific vectors of cultural extension in material and conceptual space—particular detailed configurations of the MLE with their particular performativities, and particular articulations of the electrostatic interpretive account. On the other hand, achievements of material-conceptual associations of translation marked end points, where facts could be reported and practice could (temporarily) rest. The basic image, then, is one of empirical scientific practice as the open-ended extension of existing material and conceptual culture, in which specific trajectories come to be marked out and stabilized by the achievements of associations between these elements which, as it were, transmute material performances into facts.

I turn now to emergence and posthumanism, and the first point I want to stress is that the trajectories linking the various loci of stabilization in Morpurgo's practice, and the substantive contents of those stabilizations—the material configuration of the MLE, its performance, the form of its interpretive account, a fact about nature—were temporally emergent. These trajectories and their end points were the upshot of continual revisions of modelling vectors in dialectics of resistance and accommodation, which were themselves emergent in the real time of practice. Thus neither Morpurgo nor anyone else could have foreseen the specific resistances that he would encounter in his work. That the first grain to be examined moved in the same direction when the direction of the electric field was reversed just happened in the real time of

Morpurgo's practice: likewise the appearance (twice) of a continuum of charges, and of temporally varying charges, in his later measurements. Similarly, Morpurgo's accommodations to these resistances—revising his interpretive account of his apparatus in specific ways, widening the separation between the plates, spinning the samples, the institution of new material procedures to control the magnetoelectric effect—all have to be seen as temporally emergent responses to situated obstacles. And similarly again, the success or failure of such accommodations has to be seen as becoming apparent only in practice. It just turned out that increasing the gap between the plates and so on led to findings of integral charges. As far as a real-time analysis is concerned, therefore, no preexisting principle explains or lies behind the trajectory of evolution of Morpurgo's material apparatus or its performativity, or his interpretive account of his apparatus, or the facts that he and it produced; there is only the mangle.

Now for the posthumanism of the mangle. In chapter 2, I discussed this in terms of the emergent interdefinition of human and material agency, but I have not yet said enough about conceptual practice to achieve quite such a symmetrical formulation here. In chapter 4, though, I argue that the distinct dynamics of conceptual practice should be understood in terms of what I call disciplinary agency, meaning a field of disciplined human practices. And if I can borrow in advance from that discussion to assign concepts to the human rather than the machinic realm, then I can clarify the sense in which the extension of culture in empirical practice should itself be understood as posthumanly decentered. The basic point is simple: resistances are liminal and accommodations are impartial with respect to the human/material division.

Thus, first, the resistances that Morpurgo encountered in his practice were situated *neither* in the nonhuman realm of his instrument and its performances *nor* in the human realm of concepts and conceptual structures (Morpurgo's interpretive and phenomenal accounts). Just like the resistances that Glaser encountered in the bubble-chamber project, the resistances that emerged in the Genoa experiment were situated right on the boundary of the two realms. The peculiar material performance of the first graphite grain, for example, counted as a resistance only relative to the specific set of understandings that Morpurgo brought to bear upon it (and later revised). Second, here as in chapter 2, we can see that accommodation works impartially upon both sides of the material-conceptual divide. In the face of resistance, Morpurgo tinkered *both* with the material form of his apparatus *and* with his understandings of it and the world. And the proper location of such tinkering was decided only retrospectively: the problem with the first grain could afterward be said to lie in Morpurgo's interpretive account of it, just because an amendment to the latter led to a smooth translation between the performance of the apparatus and one of Morpurgo's phenomenal models; the problem with the seventh grain (and others that followed) could be ascribed to the material instrument just because a transformation of its material configuration again led to an interactive stabilization of the material and the conceptual. As in chapter 2, therefore, we can recognize here an impure, posthuman dynamic in the extension of culture, in which the human realm of conceptual structures is important, as is the nonhuman realm of material agency, but in which neither can be regarded as a controlling center of the action. How concepts are to be extended is not determined within those concepts, but in relation to other concepts and material performances, which again contain no blueprint for their own extension—the overall process of empirical practice being oriented to the achievement of impure, material-conceptual, interactive stabilizations.

3.3 MULTIPLICITY, HETEROGENEITY, AND ASSOCIATION

> He got out his Descartes again; dipped into his Hume and Berkeley; wrestled anew with his Kant; pondered solemnly over his Hegel and Schopenhauer and Hartmann; strayed gaily away with his Greeks—all merely to ask what Unity meant, and what happened when one denied it. Apparently one never denied it.
>
> Henry Adams, *The Education of Henry Adams*

> Let us return to the story of *multiplicity*, for the creation of this substantive marks a very important moment. It was created precisely in order to escape the abstract opposition between the multiple and the one, to escape dialectics, to succeed in conceiving the multiple in the pure state, to cease treating it as a numerical fragment of a lost Unity or Totality or as the organic element of a Unity or Totality yet to come.
>
> Gilles Deleuze and Félix Guattari, *A Thousand Plateaus*

This section is both substantive and transitional. I seek to elaborate my account of the nature of scientific practice in general by assimilating my analysis of the Genoa experiments to wider themes of cultural multiplicity and heterogeneity. In the following sections of this chapter, I specialize the discussion back to specific aspects of knowledge production.

I can start by noting that, while I analyzed Morpurgo's work in terms of the interactive stabilization of just three cultural elements—his material instrument, his interpretive account of the instrument, and his phenomenal account(s) of how the world might be—there is nothing magical about the number three. Rather, we have here an exemplification of the *multiplicity* of scientific culture. Traditional accounts of science have usually been developed in terms of a unitary view of culture. The tendency in abstract discussions at least has been to characterize science in terms of one or a few theories, paradigms, research programs, conceptual nets, or whatever.[18] But while such characterizations might be adequate to the purposes for which they have been constructed—bolstering particular arguments about the objectivity or relativity of scientific knowledge, say (see chapter 6)—they are undermined in studies of scientific practice which, as I mentioned in section 1.1, confront us with the multiplicity of scientific culture, with its disunity, its patchiness and scrappiness. In Morpurgo's experiments, for example, what was clear for most of the time was the gap, the distinction between, the lack of unification of, his material instrument, his interpretive account of it (constructed in classical electrostatic theory), and his phenomenal accounts (constructed in elementary-particle theory). Far from constituting parts of any unitary culture, the difficulty of assembling these three elements into a coherent whole lay at the heart of Morpurgo's practice. And this is the second general point that studies of scientific practice bring home: a *characteristic feature* of scientific practice is that it contrives to make associations or alignments between diverse cultural elements, as, in the present instance, Morpurgo's achievement of a material performance that would translate through an interpretive account into one of his phenomenal accounts.[19]

Having introduced the theme of cultural multiplicity, I can now re-

18. The logical empiricist tradition in philosophy of science has tended to equate scientific culture with single scientific theories, Kuhn (1970) suggests a vision of scientific culture in which unitary scientific paradigms displace one another in time, Lakatos (1978) reconstructs the history of science in terms of a few competing unitary research programs, and most abstract discussions of SSK concern the dynamics of holistic conceptual nets (see, for example, Barnes 1982; and Collins 1992, chap. 6; the network model is taken from Hesse 1974, 1980).

19. "Association" is, in fact, a term of art in the actor-network oeuvre (Latour 1987); followers of Michel Foucault tend to use terms such as "alignment," "linkage," and "relay" (see the works cited in chap. 1, n. 43). The importance of seeing scientific practice as the business of making associations within a multiple and heterogeneous culture was first emphasized within the actor-network approach, although, as Gingras (1994) points out,

mark upon the related theme of cultural *heterogeneity.* While certain elements of scientific culture can be thought of as belonging to homogeneous strata—Morpurgo's interpretive and phenomenal accounts belonged, say, to the conceptual stratum of science—others are heterogeneous with respect to one another—Morpurgo's MLEs, for example, were material, not conceptual. And this observation leads us back to the posthumanism and impurity of the mangle. Just because practice moves within a heterogeneous cultural space, constructing (and deconstructing) associations and alignments between elements from differing strata, so, in general, those strata lack autonomy, contaminate one another, reciprocally structure each others' trajectories of emergence. In the present example, the material form of scientific culture and its conceptual contents grew up together in a process of interactive stabilization; more broadly, whatever axes of heterogeneity one cares to demarcate in scientific culture, the realms thus distinguished are always liable to impure mangling as associations between them are made (or broken) in practice.[20]

The idea that practice effects associations between multiple and often heterogeneous cultural elements is, then, at the heart of my overall analysis of scientific practice, not just of knowledge production. It can, for

---

this insight is often muddied in the actor-network canon by frequent descriptions of scientific culture as a "seamless web." The web is seamless only once associations have been made, and for as long as they last; the basic image should be one of multiplicity, not unitarity. Hacking (1992a) offers and exemplifies a nonexhaustive list of fifteen types of cultural element that can, in various combinations, enter into the production of scientific facts (though, coincidentally, he divides these types into three categories: "ideas," "things," and "marks"). Hacking also emphasizes the temporally emergent mutual adjustment of whatever items of his list enter into particular passages of fact production, and he expounds this position in relation to Duhem's familiar argument concerning mismatches between facts and theoretical predictions. Duhem's suggestion (1991, part 2, chap. 6) was that no particular inference follows from such mismatches; they point only to a need for adjustment somewhere in the set comprising the theory under test and all of the other theories involved in going from the first theory to an empirical prediction, and in going from a material experimental performance to an articulated fact. "No absolute principle directs this enquiry, which different physicists may conduct in very different ways without having the right to accuse one another of illogicality" (Duhem 1991, 216). Hacking argues that Duhem's suggestion needs to be extended beyond the domain of theory to include the mangling, in my terms, of the material contours of scientific culture, too. This is also the argument of Pickering 1986.

20. This means that the distinctions I make throughout the book between the material stratum of culture, the conceptual, the social, and so on, are everyday, commonsense distinctions. I do not assign them any foundational significance; I do not, for example, seek to demarcate a pure Durkheimian realm of the social. The entire thrust of my analysis, as just stated, is to show how these realms reciprocally infect each other.

example, readily be read back into chapter 2. I stressed there the interactive stabilization of heterogeneous—material and social—elements of Glaser's practice, as the size of his collaboration was both transformed in the construction of a series of material instruments and stabilized alongside the material performativity of the quenched xenon chamber. Looking ahead, my analysis of conceptual practice in chapter 4 centers on the dialectics of resistance and accommodation that arise around the production of associations in extensions of multiple elements of the conceptual stratum of scientific culture, and chapter 5 centers on relations between new material technologies and disciplined social relations in the workplace. For the remainder of this chapter, though, I will return to my discussion of knowledge production in science. Section 3.4 addresses the point that more cultural multiplicity than has so far been exemplified is typically entailed in the production of facts and in their relation to high theory. Section 3.5 is about the disciplined human practices that accompany the production of scientific knowledge—another cultural stratum, heterogeneous to the material and the conceptual. In both sections, I argue that the mangle is in play in the production of associations.[21]

3.4 REPRESENTATIONAL CHAINS

In this section, I continue to be preoccupied with the intertwining of machinic performances and scientific knowledge, and I proceed by remarking upon two respects in which the Genoa quark-search experi-

21. I should acknowledge that I cannot offer any closed definition of "association," beyond that of a bringing together of diverse elements within different projects. In the Genoa experiments, association amounted to the relation of translation between the material and the conceptual that I have been discussing; in Glaser's practice, it amounted to a relation of mutual sustenance between the material and the social; in chapter 4, it amounts to a relation of one-to-one correspondence of elements and operations in different mathematical systems; and so on. It seems to me that a closed definition of "association" would entail a finite classification of the kinds of projects in which associations were sought, and I doubt whether such a classification of scientific practice would make sense. Our general model for thinking about association might therefore be the relation between the parts of a machine when it is working: one can appreciate the fact that the parts of all sorts of working machines somehow cooperate with one another, without imagining that "cooperation" means the same thing for the components of, say, a computer, a telescope, and an atom bomb. Alternatively, I could remark that the specific kinds of association at which practice aims are always themselves open-endedly modelled on specific prior associations. Either way, I think that these remarks express, in part at least, the "irreductionist" sentiments of Latour (1988a, part 2).

ments were untypical of empirical scientific practice. Before that, however, I want to emphasize one respect in which I think they were entirely typical, namely, the centrality to them of the framing of material agency. Framing, as I defined it, is the delicate and open-ended process of reconfiguring the material culture of science in the pursuit of material performances that can be precisely aligned with conceptual structures. In the diamagnetic version of Morpurgo's experiment, the act of widening the distance between the two metal plates was the primary instance of a successful framing; in the ferromagnetic experiment, spinning the samples was another. I want here simply to suggest that framing should be understood as constitutive of empirical practice in science: part of the dynamics of the evolution of the material culture of science consists in its fine-tuning oriented to interactive stabilizations with elements of the conceptual culture of science.[22] This is the material end of the business of threading together machinic performances and scientific knowledge; the latter in general terminates not just in captures but in framings of material agency. Now I want to think about the realm of concepts and articulated knowledge in more detail, and I begin with the untypical aspects of the Genoa experiments.

The first oddity of the Genoa experiments concerns the fixity of Morpurgo's pair of phenomenal accounts. Morpurgo's willingness to accept only one of a disjoint pair of well-defined outcomes is what makes his experiments a perspicuous site for study. It throws into sharp relief the mangling of Morpurgo's material instrument and his interpretive account of it, and these are the hardest aspects of empirical practice

22. Many examples of framing in material practice come to mind besides the two just mentioned. See Holton 1978 on the evolution of the material form of Millikan's oil-drop experiment (on which Morpurgo modelled his own work), Fleck [1935] 1979 on the temporal evolution of the material techniques of the Wassermann reaction, Shapin 1979 on the adoption of different material techniques (in relation to different conceptual systems) by the phrenologists and their opponents, Trenn 1986 on the drastic material reconfiguration of the Geiger-Müller counter (in relation to changing beliefs about cosmic rays), and Galison 1983 and Pickering 1984a on the transformations of the Harvard-Pennsylvania-Wisconsin-Fermilab neutrino detector implicit in their confirmation of the discovery of the weak neutral current. See also the other historical studies discussed in Galison 1987. Gooding's studies (1990, chap. 5; 1992) of Michael Faraday's practice leading up to his construction of the first electric motor are wonderful exemplifications of the simultaneous capture, framing, and representing ("construal") of material agency (though the irony, from my perspective, is that when Gooding speaks of "putting agency back into observation," the agency he has in mind is purely human; remaining faithful to the representational idiom, Gooding even speaks of Faraday's first construction of the motor as an "observation" [1992, 91]).

in science to grasp (at least from a traditional philosophy-of-science perspective). More usually, though, the specific forms of phenomenal accounts are themselves at stake in experiment. The point of many experiments in physics, for instance, is to fix the value of some continuously variable theoretical parameter. The interactive stabilizations that translate the material into the conceptual should therefore be seen as being made in the mangling of *all* of the cultural elements that enter into them. My claim is certainly *not* that phenomenal accounts have any necessary priority and that other cultural elements are necessarily molded around them (see the discussion of top-down relativism in section 6.5 for more on this theme). It is worth noticing, for example, that the space of phenomenal accounts in which Morpurgo worked had itself been dramatically reconfigured just prior to that work—his quark searches would have been inconceivable prior to Gell-Mann's and Zweig's quark proposal—and that nothing guaranteed in advance that similarly striking changes might not have been subsequently worked by Morpurgo (or someone else). Indeed, one response to the controversy that enmeshed the later stages of the Genoa experiment (section 6.6) took the form of tinkering with theoretical understandings of the properties of isolated quarks. I have no complete explanation of why Morpurgo himself remained faithful to his pair of phenomenal accounts even when his findings seemed to favor neither, but to make sense of this, one would no doubt need to think about the contingent fact that, as it happened, he did manage to achieve interactive stabilizations that preserved them; about his focus on his particular experiment rather than any theory of continuous charges, of which, as it also happened, none were on offer in the physics of his day; about particular design features of the MLE; and so on.[23]

Now for the second oddity of the Genoa experiments. These experiments were unusual in the close relation between their findings and high theory. On the face of it, the facts that Morpurgo produced directly ruled out any version of quark theory that suggested that quarks could be found in isolation (though, as mentioned in section 6.6, matters were

23. I am told by Professor Morpurgo (letters, 22 and 25 November 1990) that he felt that his apparatus was capable of detecting subelectrons, particles having very small charges, neither integral nor third-integral, but that this capacity resided in the measurements of step lengths, not residual charges (see n. 7 above); and, as it happened, no evidence for subelectrons showed up in the step-length measurements. My last conversation with Professor Fairbank before his death also touched upon the possibilities for finding charges that were neither integral nor third-integral in his quark-search experiment.

not quite so simple). The relation between experiments, facts, and theory in science is rarely so immediate, however.[24] More typically, complex *representational chains,* as I call them, lead upward from the realm of instruments and material performances through successive layers of abstraction. At stake here, then, is the idea that the conceptual stratum of scientific culture is itself multiple rather than monolithic, and that many disparate layers of conceptualizations, models, approximation techniques, and so on have typically to be linked together in bringing experiment into relation with the higher levels of theory. In his study of solar-neutrino astronomy, for instance, Pinch (1985, 1986) documents the stages of processing that led from happenings in a tank of dry-cleaning fluid, through instrumental traces on paper, to records of nuclear transformations, to accounts of neutrino-induced events, to measures of the intensity of the neutrino flux from the sun striking the earth; and he describes the other stages of processing that led from general theories of particle and nuclear interactions through models of stellar dynamics and specific solar parameters to predictions of that same neutrino flux.[25]

The question thus arises, of course, of how such representational chains and their dynamics fit in with my analysis. I do not think that they pose any special problem. I have no detailed examples to discuss, but my suggestion is that they are produced, extended, elaborated, connected to one another (perhaps at the expense of dismantling other crosscutting chains), in just the way we have already been discussing. All of the elements of representational chains, whatever they are, are subject

24. In the directness of their relation to theory and in their aspect of crucial experiments making choices between disjoint possibilities, Morpurgo's experiments conform nicely to the "theory-testing" image traditional in philosophy of science. By describing these aspects of Morpurgo's as untypical, I want, among other things, to assert that much scientific empirical and observational work does not have this theory-testing quality. See, for example, Hacking 1983 on the many purposes that experiment can serve, and the examples discussed in Gooding, Pinch, and Schaffer 1989. In Pickering 1984b, chapter 11, I argued at length that, in the history of elementary-particle physics, the relation between QCD (the gauge theory of the strong interactions) and its associated experimentation did not conform to the theory-testing image.

25. See also Shapere 1982 on this example. The essays collected in Lynch and Woolgar 1990 variously explore representational chains spanning multiple cultural elements and sequences. Lynch's discussion (1991b) of the computer production of digitized images from numerical data is also relevant here. Latour (1987, 232–47) offers a general image of representational chains as "cascades" of "re-representation" in which statistical operations and mathematical formalisms interconnect *n* levels of abstraction; Cartwright (1983) is especially concerned with connections between high-level laws and phenomenological accounts in science.

to open-ended extension in practice and to interactive stabilization inasmuch as they function (or might function in the future) as parts of such chains. They are, that is, extended in mangling, just like Morpurgo's material instrument and his conceptual accounts. Ian Hacking's commentary on Nancy Cartwright's work can serve to indicate what is at stake here. As Hacking puts it:

> The relations of models to theory and to phenomena are various and complex. Approximations seem more straightforward. Cartwright shows that they are not. Our usual idea of an approximation is that we start with something true, and, to avoid mess, write down an equation that is only approximately true. But although there are such approximations *away* from the truth, there are far more approximations *towards* the truth. In many a theory of mathematical physics we have a structural representation with some equations at a purely hypothetical level, equations which are already simplifications of equations which cannot be solved. In order to make these fit some level of phenomenological law, there are endless possible approximations. After a good deal of fiddling someone sees that one approximation tallies nicely with the phenomena. Nothing in the theory says that this is the approximation we shall use. Nothing in the theory says that it is the truth. But it is the truth, if anything is. (1983, 218)

I am not sure about the concluding remarks on truth, but otherwise this is a nice description of stabilization at the higher levels of representational chains, in which appropriately contrived approximation schemes serve, like Morpurgo's interpretive accounts, as translation operators, but this time linking different levels of conceptual abstraction and cementing associations between theories and "phenomenological laws." I would only add three points. First, the stabilizations in question should in general be seen as interactive ones, in recognition that the specific contents of theories and laws as well as those of approximation schemes are subject to open-ended development in the production of associations and links up and down such chains.[26] Second, the produc-

26. "Theories," "laws," and "approximations" are traditional philosophy of physics talk, but my point here is a general one. Lynch, for example, offers a relevant discussion of the computer production of digitized images from numerical data: "A sense of what the picture shows guides the project [of image production and enhancement]: a sense of how many pixels the object's profile should cover; what intensities and intensity gradients are appropriate for a 'point source' versus a nebular object; and what order of symmetry and asymmetry should be expected" (1991b, 70). In my terms, Lynch gestures here toward the interactive stabilization of (1) a set of operations upon a set of numerical data, (2) the image/object generated by those operations, and (3) phenomenal accounts (theory) of that

tion of linkages along representational chains should not be thought of as trivial. Nothing guarantees in advance that any given link can be achieved as intended. Pinch's study of solar-neutrino physics can again stand as our example. The chains leading upward from experiment and downward from models of neutrino production in the sun have never been joined up very well—far fewer neutrinos are experimentally detected than are predicted. This "solar-neutrino problem"—a resistance to the alignment of representational chains, in my terminology—is well recognized in physics and has been the focus of much experimental and theoretical work. And, in general, I think that much scientific practice should be understood as oriented to the accommodation of such resistances—to the improvement of alignments along representational chains.

My third point follows on from this. As already stated, I have not attempted in this chapter to give an analysis of how dialectics of resistance and accommodation can arise in such conceptual practice; that is the topic of chapter 4. Presuming once again upon that chapter, though, I can conclude this section by saying that scientific knowledge—from the realms of high theory to the humble domain of empirical facts—should be understood in terms of representational chains ascending and descending through layers of conceptual multiplicity and terminating in captures and framings of material agency, with the substance and alignments of all the elements in these chains formed in mangling. Now we can turn from the multiplicity of the conceptual stratum of scientific culture to another axis of cultural heterogeneity.

3.5 DISCIPLINE

Section 3.3 thematized the heterogeneity of scientific culture by focusing on the manglings and interactive stabilizations involved in linking its

object (which are themselves modified in the process). One can make the overall picture of representation in science more complicated by referring to another of Lynch's studies, this time on "rendering practices." Lynch (1985b) points to the progressive material transformations worked upon rats—from living being, to dissected brain, to hippocampus, to slices thereof, to electron micrographs—in the production of knowledge on "axon sprouting." These practices are intended to "upgrade" and "make docile" (Lynch's terms) or frame (mine) rats for representational processing, and they move the rats along *prerepresentational* chains, as one might say. One should see these chains and the rendering practices that act as translation operators along their length as also mangled in practice. Knorr-Cetina (1992) discusses similar prerepresentational material transformations in a molecular-biology laboratory and makes interesting contrasts with other disciplines.

material and conceptual strata. This same theme can be pursued along many axes, and to conclude this chapter, I want to focus on just one of them, the mangling of human agency that goes along with captures and framings of material agency in the production of scientific knowledge. This topic has already been broached in the discussions of the mangling of human intent in section 2.3 and of the mangling of the scale of human actors and their social relations in section 2.4. Here, though, I pick up it up from a fresh angle: I want to talk about the disciplining of human agency.[27]

To explain what is at stake, I can reiterate the argument I made in section 1.3 about the relation between machines and the repetitive human performances that surround them, set them in motion, and channel them. My suggestion there was that a reciprocal tuning is at work in scientific practice, which simultaneously delineates the material contours of machines and their performances and the regularized human actions that accompany them. Or, to put it another way, that the open-ended dance of agency that is scientific practice becomes effectively frozen at moments of interactive stabilization into a relatively fixed cultural *choreography*, encompassing, on the one side, captures and framings of material agency, and, on the other, regularized, routinized, standardized, disciplined human practices.[28] I think such choreographies are omnipresent in all of our dealings with machines, but in this section I want

27. In finding it necessary to examine the disciplining of human agency in science, I seem to part company with Hacking, who is keen to keep the human and social dimensions out of an analysis of practice that looks otherwise like mine: "Those tired words 'internal' and 'external' seem useful here," he writes; "I have been offering a taxonomy of elements internal to an experiment" (1992a, 51). One can, of course, proceed in this way, and the preceding sections of this chapter do so; what puzzles me is Hacking's reluctance to acknowledge that this is only part of the analysis of empirical practice in science. Quoting a list of dimensions of scientific culture that I once gave, Hacking deletes "forms of life" from it, saying "I do not omit [this] by inadvertence," but he does not offer reasons (1992a, 31). I return to the concept of a form of life below (n. 43).

28. Recall the distinction between practice and practices made in section 1.1; my argument in this section is that *practices* are interactively stabilized together with other cultural elements *in practice*. The early literature on SSK tended to speak of "skills" and "tacit knowledge" as a way of emphasizing the detailed and embodied constitutive involvement of human beings in the production and extension of scientific knowledge (see, for example, Collins 1992; Collins's target was the usual one for SSK, namely, traditional philosophical thought that sought to deny any important involvement of the human in science beyond an idealized rationality). Skill talk, however, is slippery. Since skills are, more or less by definition, invisible, one tends to find oneself talking about skill tokens rather than skill itself—about the objects, say, to which particular skills relate: "Morpurgo was very

to remain with the problematic of knowledge and to talk about disciplines as they enter into fact production.

I proceed, as usual, by example, but I leave quark searches behind here for a better documented account that, I concede in advance, is in one respect peculiar. I am going to discuss not the first-time performance of a genuine scientific experiment but the attempted reperformance (reconstruction) of a historically important experiment. The reason for my choice is that the study to be discussed is the most perceptive that I know on the topic at issue, no doubt because the author of the study was explicitly and self-consciously interested in questions of discipline (unlike scientists themselves). I suggest later that the peculiarity of my example does not undermine the generality of the conclusions that I draw from it.[29]

---

skilled in Millikan-type charge measurements," and so forth. As we shall see, while it seems to be appropriate to incorporate the idea that specific skills are components of discipline, my notion of discipline is broader than that of skill, and seeks to foreground more visible—inspectable and documentable—features of human agency. Connected perhaps with the slipperiness of skills and tacit knowledge is a tendency to attribute odd properties to them. For example, in a typical locution, Collins remarks of one of his case studies, "It is clear that there were long periods when, though he did not have laser building ability, *H did not know* that he did not have it except by reference to the fact that the laser did not work" (1992, 62). This notion of "laser building ability" (conceived by Collins to be largely tacit) reminds me of the frequently expressed opinion among scientists that it is articulated knowledge (theory) that "enables" them to build machines. My argument throughout this book is that *nothing*—in Collins's or the scientists' sense—"enables" us to build machines: when we succeed in doing so, it is via a fortunate passage through the mangle, the disciplinary aspects of which are discussed below. Collins's relocation of the magical properties ascribed by scientists to scientific theory to the tacit realm of human skills amounts to an unwarranted puffing up of human agency that is of a piece with his consistent humanism.

29. Besides the reperformance of actual scientific experiments, I can think of two other ways to investigate empirically questions of discipline. One is historically, using conventional textual methods, though this can be difficult for the reason just stated: the accretion of human disciplines in science tends to leave little documentary trace. Nevertheless, I outline some of the practices established in Morpurgo's quark experiment in note 37 below. The other angle of attack is ethnographic, and here the ethnomethodological studies of Michael Lynch and collaborators are wonderfully insightful on the topic of discipline in general (Lynch 1985a, 1985b, 1991b, 1992a, 1992b; Garfinkel, Lynch, and Livingston 1981; Lynch, Livingston, and Garfinkel 1983). Unfortunately, from my point of view, Lynch seems relatively uninterested in cultural change through time. In this respect, ethnomethodology mirrors sociology's traditional preoccupation with questions of cultural reproduction rather than change, and one consequence is that Lynch's studies reveal little about the mangling of practices. Conversely, I think that Lynch's very important Wittgensteinian discussion (1992a, 1992b) of an "internal relation" between the human, material, conceptual, spatial, and so forth, components of scientific practices would be

My text in what follows is Heinz Otto Sibum's account (1992) of his attempts to repeat, as closely as possible, measurements of the mechanical equivalent of heat reported in 1850 in the *Philosophical Transactions* by James Prescott Joule.[30] Figure 3.9 reproduces Joule's diagram of his apparatus. Its principle is straightforward. The two weights at the extremes of the apparatus are allowed to fall, and the threads to which they are attached drive a paddle wheel between baffles in a central container full of water. The weights are then winched back to their starting positions, and the process is repeated many times. The temperature of the water is measured at the start and finish of the experiment, and thus, from a knowledge of how far the weights have fallen and how many times, the constant of proportionality between a given quantity of mechanical work and the heat that it turns into can be immediately calculated (taking into account estimates of various corrections, due to heat losses through radiation, mechanical energy losses due to friction, and so on).

Sibum describes his attempts to reconstruct Joule's work as—in my terms—a fascinating dialectic of resistance and accommodation in his "real time performance" (3). Having built a replica as close to Joule's apparatus as he could, Sibum notes that he first set out to perform the experiment with the collaboration of a graduate student, Peter Heering. The idea was that they could divide the labor of winding the weights and making the required measurements. Immediately a problem—quite unremarked in Joule's account—became apparent. Whenever Sibum and Heering entered the room in which the experiment was to be performed—a modern air-conditioned laboratory—the temperature began to fluctuate beyond the bounds they could allow if they were to make measurements of an accuracy comparable to Joule's. Their accommodation to this resistance was to make two procedural rules: (1) allow one hour after entering the room for the temperature to stabilize before starting the experiment, and (2) exclude witnesses from the room (whose presence would further imperil temperature control). They also found problems in using the very sensitive thermometers that, copying Joule, they were using. Again, they formulated rules to deal with this:

---

much clearer if the time dimension of practice were recognized there. As I understand it, the internal relation that Lynch discerns *is* the set of associations and alignments of heterogeneous cultural elements and strata that is the product of repeated impure manglings over time.

30. I thank Otto for valuable discussions. Gooding (1989) discusses attempted repetitions of Faraday's work along similar but less detailed lines.

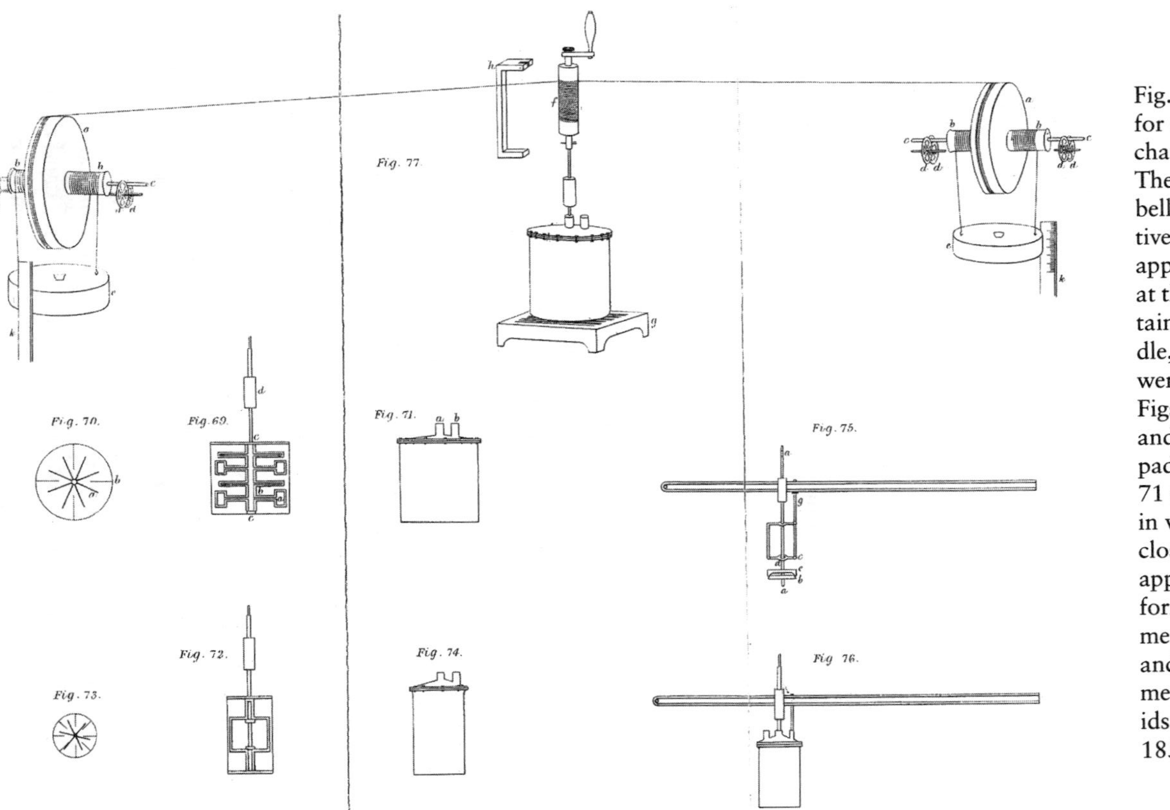

Fig. 3.9. Joule's apparatus for determining the mechanical equivalent of heat. The drawing at the top labelled "fig. 77" is a perspective view of the complete apparatus, with the pulleys at the extremes and the container of water in the middle, in which the paddles were driven by the roller *f*. Figs. 69 and 70 are vertical and horizontal plans of the paddles and baffles, and fig. 71 shows the copper vessel in which they were enclosed. Figs. 72–74 show apparatus used for performing similar experiments on mercury; figs. 75 and 76 were for experiments on the friction of solids. Reprinted from Joule 1850, plate 2.

"The act of reading the thermometer requires a certain technique which includes the right timing for taking measurements. The radiation effect between the actor and the instrument was very difficult to master. Long experience in using the particular thermometer was needed. We helped ourselves first by creating rules such as: wait two minutes then read off, and continue reading every minute. Then take the mean temperature" (4).

Other resistances became apparent. On Joule's account, a weight of 26 kilograms had to be wound up a distance of 1.4 meters twenty times in 35 minutes. "We failed completely, simply on physical grounds. We had to share the job because of lack of strength. Even then, it took us one hour and twenty minutes to perform successfully" (4). The lack of strength of Sibum and Heering created yet another problem: the falling weights heated the water by half a degree Fahrenheit, but their sweating increased the room temperature by a whole degree—their bodily exertions thus proved to be a major source of disturbance, getting in the way of accurate observation. As Sibum puts it: "An athlete's physical condition would prevent unnecessary temperature increases in the room during the trial . . . Probably an unknown brewing mate [i.e. someone strong and fit] was hired by Joule to do the job" (28). Another obstacle to this attempt to track Joule was not so much a resistance to actually doing the experiment, as a resistance to thinking of it as a reconstruction of Joule's. The speed of fall of the weights "varied markedly" from that recorded by Joule. Sibum's accommodation to this resistance was to suspect that there was something amiss with the material form of his apparatus, and to explore this possibility he examined examples of Joule's actual apparatus held at the Museum of Science and Industry at Manchester and at the Science Museum in London. He took measurements of the latter and built a replica, which he used in a second series of experiments, this time working alone (perhaps he had got into shape by then). With this new replica, the weights fell at a speed "very close" to that reported by Joule (8), pointing to the conclusion that the apparatus described by Joule in his published account was not the one on which he had made the reported measurements.

Sibum also made another change in this second reconstruction. In search of a stable temperature environment, and paralleling a similar move by Joule from his laboratory to a brewing cellar, Sibum moved his apparatus into "an old powder tower in Oldenburg which was used to store meat during the eighteenth century. It was spacious and the material construction allowed me to expect that its temperature conditions would give further information about Joule's work place" (6). Despite the stability of this environment in comparison to a modern laboratory, however, Sibum found that he had still to allow a period of acclimatization after entering the room before starting his experiments. Further: "Difficulties also arose through placing thermometers in the room: temperature differences occurred through movements of the measuring devices. Practicing with the thermometers in the room over weeks already improved my abilities as I accustomed myself to the room and its conditions" (7).

Other problems arose with thermometry, too, and were solved as best Sibum could: "Ending the experiment also needs a lot of experience. The correct timing and the appropriate habit of taking the temperature measures is absolutely crucial. Presumably, a run should cease when the final, stable, temperature has been reached. The Beckmann thermometer [the nearest equivalent to Joule's lost originals that Sibum could find] indicated a stable temperature within six minutes after putting them into the water" (7). He also records that this version of the experiment required a practiced deftness and precision of touch: "I had to maintain absolutely smooth revolving pulleys in order to guarantee a secure drive for the delicate paddle-wheel . . . No disturbance in the machinery should occur otherwise the paddle-wheel as well as the thermometers could be destroyed. Perfect mastery of this part of the performance was reached when no swinging of the system occurred. The fragile design of the vessel and axle avoided a major source of error [friction]. But it made necessary an 'artistic' mechanical performance" (9).[31] In the end, Sibum got a value for the mechanical equivalent of heat within two standard deviations of Joule's result and the value accepted today.[32]

Here, then, we have a clear and detailed account of the disciplining of human performance, of the delineation of the routinized contours of human action implicit in the capture and framing of material agency and the production of facts.[33] Sibum learned, in real-time practice, that

31. Sibum also discovered that "[t]he handle on top of the axle was much too short in order to wind up the weights. The Science Museum model's handle couldn't be the one which he [Joule] or anyone else used during the experiment. In that shape it could only have been used as a demonstration device" (9).

32. Sibum's result was 746.89 ft. lbs./BTU ± 2.1%, Joule's was 772.692, and the modern value is 776.1 (Sibum 1992, 10).

33. There is more to Sibum's essay than I have explained in the text. Having discovered the difficulty of thermometry involved in Joule's experiment, Sibum traced Joule's competence in the required techniques back to contemporary developments surrounding the rationalization and industrialization of brewing. Joule belonged to an important brewing

performance of the experiment required uninterrupted solitude, physical strength, a practiced delicacy of touch, a period of rest and acclimatization before experiments were begun, a particular positioning of the thermometers, a certain timing of thermometric measurements, and so on.[34] None of these requirements could have been envisaged in their specifics in advance of Sibum's actual experimental work. They were the upshot of a temporally emergent dialectic of resistance and accommodation—of problems arising and being addressed in practice—and together they constituted, as it were, a temporally emergent, autochthonous discipline of human agency: Sibum found that he had to conduct himself in these specific ways if he were to reperform Joule's experiment. And, further, this mangling of Sibum's performances had, as usual, a posthuman character. On the one hand, Sibum's human performances were evidently stabilized against the captures and framings of material agency first achieved by Joule. On the other, the material form of Sibum's apparatus and its specific performativity were themselves at stake in these reconstructions: Sibum found that he had to discard a central component of his original apparatus and replace it with a different version if he were successfully to shadow Joule. The disciplined character of Sibum's human performance, then, was *interactively stabilized* alongside the material (and conceptual) elements with which he worked.[35]

And this is the general point that I want to make: the disciplined human performances that accompany machines and their insertion into

---

family and had himself been involved in these developments. The brewing industry was thus a surface of emergence for Joule's scientific work. Sibum indicates that the community of precision-instrument makers was another surface of emergence: nearly all of Joule's scientific equipment was built by a Manchester instrument maker, John Benjamin Dancer. One can make a direct connection here to the "performative historiography" that I discuss in section 7.3.

34. Sibum groups the items in this list under the heading of "gestures" and speaks of "gestural knowledge" rather than "discipline." Not much is at stake in the different terminologies, but "gesture" seems too active a word for some of the items in question: being alone, being strong, doing nothing. Likewise, reconstructing Joule's experiment in a powder tower rather than a laboratory is not a "gesture" as I would use the word. Joule's and Sibum's shifts in their places of experiment point, rather, I think, to the heterogeneous and emergent mangling of the *physical architecture* of science. For more on this topic, see Hannaway 1986; and Shapin 1988b.

35. Collins and Kusch (forthcoming a) discuss their attempts to operate a modern vacuum pump, and nicely exemplify the stabilization of human performance against captures of material agency, though, since they had to take the pump as they found it, the stabilization here was not an interactive one—Collins and Kusch were transformed (disciplined) by and in their experience, but the material culture of science was not.

representational chains are themselves the products of the mangle, the products of temporally emergent sequences of resistance and accommodation, interactively stabilized in conjunction with elements of all sorts of heterogeneous cultural strata—material, conceptual, social (to borrow from the conclusions of chapter 2), and whatever. One might worry, as I mentioned earlier, about the peculiarity of my example—about whether an attempted reconstruction of a historical experiment can stand as a representative for genuine experimental practice in science. But it seems to me that such worries are misplaced. While Joule's practice must have involved more manglings than Sibum's (at least the latter knew pretty well what value for the mechanical equivalent of heat he was aiming at), it seems impossible to doubt that much the same kind of manglings bore upon Joule's human performance as bore upon Sibum's.[36] We are, then, entitled to take Sibum's record as exemplary of empirical science itself.[37]

To conclude this discussion of discipline, I want to mention that there are many interesting and important lines along which one might carry the present line of thought forward—and along which, indeed, other scholars have carried it. Sibum's example, as I have said, foregrounds the autochthonous discipline proper to particular experiments in science, painstakingly fashioned in extended passages of scientific practice. Any comprehensive discussion, however, would need to cover the *detachment* of disciplines from particular sites. Michael Lynch's studies, for

36. To pursue this worry a little further, one might wonder whether the discipline that emerged in Sibum's practice was exactly the same as Joule's. Again I suspect that this is not a genuine worry at all. As Jordan and Lynch (1992) show in their study of the "plasmid prep"—a standard technique in molecular biology—the autochthonous disciplines of laboratory practice typically do vary from one practitioner to another. What counts is the interactive stabilization of human and material performances to create a cyborglike machine adequate to some purpose.

37. I return here briefly to the quark-search example, and note some parallels with Sibum's account. Thus, in the early diamagnetic experiments, Morpurgo learned in practice that he had to be careful not to touch the knobs that adjusted the separation of the plates during a measurement: "it was sufficient to touch [a] knob of one of the platelets to see, after a while, the grain moving due to the temperature gradient which was created in this way" (1972, 5). I caused a certain amount of anguish when I picked up the central component of the ferromagnetic experiment (now dismantled): Professor Morpurgo informed me that touching the plates used to apply the electric field caused "enormous patches" (see the discussion above of the patch or Volta effect). Marinelli and Morpurgo (1982, 213–15) offer "a sort of operation manual" for their ferromagnetic levitation electrometer, consisting of twenty-two detailed sequential steps involved in a typical experimental run, eighteen of which have to be performed before measurements begin.

example, explore what one might call baseline practices of representation, deployed in many situations and particular to none.[38] At a greater level of detachment, it would be very interesting to study the transmission of disciplines in scientific training and apprenticeship, though, as far as I am aware, not a great deal of detailed research has been devoted to this topic.[39] In chapter 4, I invoke such detached disciplines as part of my explanation of conceptual practice, and I analyze their mangling in practice.

Another angle on the detachment of disciplines from particular sites concerns the construction and imposition of explicit disciplinary regimes to orchestrate the mass production of scientific knowledge by socially extended collectives.[40] Thus, to return to an example from chapter 2, Galison (1985) documents the various disciplines imposed upon skilled and unskilled labor—physicists and scanners—in the operation of Alvarez's 72-inch bubble chamber and in the conversion of its photographic product into articulated knowledge. And Schaffer's study (1988) of developments in the production of astronomical knowledge during the nineteenth century makes it clear that such disciplines did not originate in the post–World War II era that we traditionally associate with big science.[41] I return to a consideration of disciplinary regimes (in the

38. Thus Lynch (1985b) takes the specific scientific project of research into axon sprouting as an occasion to discuss general strategies of rendering material objects (rats' brains) into a Cartesian graphic space, and he (1991b) discusses two "topical contextures" that he calls "opticism" and "digitism" involved in scientific (and artistic) representation. The first accretes around traditional understandings of geometrical optics; the second is the regime of supercomputers and digital image processing.

39. Examples that come to mind are Lynch, Livingston, and Garfinkel's commentary (1983) on laboratory experiments in introductory chemistry; Hartley's analysis (n.d.) of pedagogical texts on the "touch" in landscape painting; and Graham's study (1988) of postgraduate training in sociology (though Graham focuses primarily on conceptual rather than material practice). See also the works cited in Geison 1993. An interesting theme that emerges from such studies concerns the mangling of students' practices around pedagogical texts (manuals and so on) that are treated as at once authoritative fixed points and in need of open-ended interpretation (which itself gets fixed, if at all, in practice).

40. There are obvious connections to be made here with the work of Michel Foucault (especially with Foucault 1979; see also the Foucauldian studies of the history of the social sciences cited in chap. 1, n. 43). Lynch (1991b) elaborates on the alignments that can be made between Foucault's thought on discipline, ethnomethodological analyses of practices, and actor-network theory.

41. Interestingly, both Galison and Schaffer document the use of additional layers of machines, or new uses of existing machines, in the disciplining, calibration, and standardization of the scientific labor force. Thus the introduction of the chronoscope was crucial to the disciplining of astronomical observation in the nineteenth century—"Artificial stars and galvanic clocks substituted eye-and-ear methods" (Schaffer 1988, 118)—while the

factory) and their mangling in chapter 5. Finally, in this connection, I can mention what Gooding (1992, 105) calls, somewhat misleadingly, the "packaging of skills"—meaning the pursuit of particular material configurations that effect some specified capture of material agency while minimizing the sensitivity of the instrument or machine in question to the precise contours of the human agency that envelops it.[42] All of these aspects of the disciplining of human performativity should, I think, be seen as emergent products of heterogeneous mangling.

Where have we got to? The overall view of scientific culture developed so far is one of multiplicity and heterogeneity spanning all sorts of machines and instruments, conceptual and representational structures and systems, disciplined practices, social actors of different scales and their social relations (chapter 2), and so on. My specific concern in this chapter has been with how knowledge fits into this picture, and my suggestion is that we should see empirical scientific knowledge as constituted and brought into relation with theory via representational chains linking multiple layers of conceptual culture, terminating in the heterogeneous realm of captures and framings of material agency, and sustaining and sustained by another heterogeneous realm, that of disciplined human practices and performances. This is my basic angle on how to think about representation in the performative idiom. Beyond this, I have offered an analysis of empirical practice in science, of the production of new knowledge, in terms of the open-ended extension of all the cultural elements just mentioned. The proper vectors of such extensions—the precise configurations of machines and instruments and their associated material performances, the precise substance of representational chains and facts, the precise contours of disciplined human agency—are determined in practice, I have argued, in the achievement of interactive stabilizations among them. Nothing substantive in scientific culture

very computers that Alvarez's scanners used in making measurements on particle tracks were programmed to lead the scanners through their steps in a dialogic dance and, if necessary, call them to account, asking for scanners to repeat operations if the computer input did not seem to make physical sense (Galison 1985, 343). In science as in industry, human discipline has evolved hand in hand with metrology and the standardization of machines—on which topics, see Schaffer 1992a; O'Connell 1993; and Latour's seminal discussion, 1987, 247–57.

42. Gooding's example concerns Faraday and the electric motor. Having arrived at the motor via a tortuous sequence of material, conceptual, and disciplinary manglings, Faraday built a small demonstration device, which he distributed along with a simple list of instructions for its assembly and use (Gooding 1992, 105–6).

or anywhere else—this is my claim, and this is what fascinates me—necessarily endures through and explains the process of cultural extension; everything in scientific culture is itself at stake in practice; there is nothing concrete to hang onto there. But still, I have tried to show that one can grasp this process and make sense of it in terms of the mangle—the temporally emergent dialectics of resistance and accommodation that inevitably arise around the association of multiple cultural elements. It is the mangle that determines in time what scientific machines will look like, what scientists will believe, how they will conduct themselves around those machines, and how (if at all) these pieces and others will hang together and relate to one another.[43] But one piece of my own analysis is still missing: it is time to talk about conceptual practice.

43. Something like this image of a bloc of interactively stabilized heterogeneous elements is often referred to in the science-studies literature as a "form of life" (invoking Wittgenstein 1953). See, for example, Collins 1992; Shapin and Schaffer 1985; and Pickering 1989a. Alternative designations that avoid the apparent human-centeredness of this locution include "actor-network" (under the proviso that "actor" refers indifferently to humans and nonhumans), the "cyborg" imagery of Donna Haraway (1991b), and Lynch's "topical contextures" (see n. 38 above).

FOUR

Concepts

Constructing Quaternions

Similarly, by surrounding $\sqrt{-1}$ by talk about vectors, it sounds quite natural to talk of a thing whose square is -1. That which at first seemed out of the question, if you surround it by the right kind of intermediate cases, becomes the most natural thing possible.

Ludwig Wittgenstein, *Lectures on the Foundations of Mathematics*

How can the workings of the mind lead the mind itself into problems? . . . How can the mind, by methodical research, furnish itself with difficult problems to solve?

This happens whenever a definite method meets its own limit (and this happens, of course, to a certain extent, by chance).

Simone Weil, *Lectures on Philosophy*

Something is missing from chapter 3. An asymmetry exists: machines are located in a field of agency but concepts are not. Thus, while it is easy to appreciate that dialectics of resistance and accommodation can arise in our dealings with machines—I have argued already that the contours of material agency emerge only in practice—it is hard to see how the same could be said of our dealings with concepts. And this being the case, the question arises of why concepts are not mere putty in our hands. Just why did Morpurgo, for instance, have to work through a genuine and uncertain dialectic of resistance and accommodation in suturing together the performance of his material apparatus and his interpretive and phenomenal conceptual accounts? Why was his work of fact production not trivial? Why did he not just tailor his interpretive account of his apparatus to produce, say, findings of third-integral charge (which might well have put him in line for a Nobel Prize)? Or why did he not just let his phenomenal accounts conform to what seemed usually to follow from his experiments, namely, the existence of a continuous distribution of charges? More generally, why are the kinds of alignments in extended representational chains that I discussed in section 3.4 nontrivial achievements? Why is conceptual practice difficult? "How can the workings of the mind lead the mind itself into problems?" It seems to me that one cannot claim to have an analysis of scientific practice until one can suggest answers to questions like these. In this chapter I argue, first in the abstract then via an example, that the mangle is again what we need.[1]

4.1 DISCIPLINARY AGENCY

When we think, we are conscious that a connection between feelings is determined by a general rule, we are aware of being governed by a habit. Intellectual power is nothing but facility in taking habits and in following them in cases essentially analogous to, but in non-essentials widely remote from, the normal cases of connections of feelings under which those habits were formed.

Charles Sanders Peirce, "The Architecture of Theories"

The student of mathematics often finds it hard to throw off the uncomfortable feeling that his science, in the person of his pencil, surpasses him in intelligence.

Ernst Mach, quoted by Ernest Nagel, *Teleology Revisited*

My analysis of conceptual practice depends upon and elaborates three ideas that have already been introduced: that cultural practices (in the plural) are disciplined and machinelike; that practice, as cultural extension, is centrally a process of open-ended modelling; and that modelling takes place in a field of cultural multiplicity and is oriented to the production of associations between diverse cultural elements. I can take these ideas in turn.

1. Latour is right to complain about the dearth of empirical studies and analyses of conceptual practice in science: "almost no one," as he puts it, "has had the courage to do a careful anthropological study" (1987, 246). Whether failure of nerve is quite the problem, I am less sure. Much of the emphasis on the material dimension of science in recent science studies must be, in part, a reaction against the theory-obsessed character of earlier history and philosophy of science. In any event, Livingston's ethnomethodological explorations of mathematics (1986) are a counterexample to Latour's claim, and the analysis of conceptual practice that follows is a direct extension of my earlier analysis of the centrality of modelling to theory development in elementary-particle physics (Pickering 1981c, 1984b; and see the literature on metaphor and analogy cited in chap. 1, n. 30). Nevertheless, resistance and accommodation are not thematized in my earlier analyses, and it might be that this exemplifies the lack of which Latour complains.

Think of an established conceptual practice—elementary algebra, say. To know algebra is to recognize a set of characteristic symbols *and how to use them.* As Wittgenstein put it: "Every sign *by itself* seems dead. *What* gives it life?—In use it is *alive*" (1953, sec. 432, quoted in Lynch 1992b, 289). And such uses are structured by what I called detached disciplines in chapter 3; they are machinelike actions, in Harry Collins's terminology. Just as in arithmetic one completes "3 + 4 =" by writing "7" without hesitation, so in algebra one automatically multiplies out "$a(b + c)$" as "$ab + ac$." Conceptual systems, then, hang together with specific disciplined patterns of human agency, particular routinized ways of connecting marks and symbols with one another.[2] Such disciplines—acquired in training and refined in use—carry human conceptual practices along, as it were, independently of individual wishes and intents. The scientist is, in this sense, passive in disciplined conceptual practice. This is a central point in what follows, and to mark it and to symmetrize the formulation, I want to redescribe this human passivity in terms of a notion of *disciplinary agency.* It is, I shall say, the agency of a discipline—elementary algebra, for example—that leads us through a series of manipulations within an established conceptual system.[3]

I will return to disciplinary agency in a moment, but now we can turn from disciplined practices to the practice of cultural extension. A point that I take to be established about conceptual practice is that it proceeds through a process of modelling. Just as new machines are modelled on old ones, so are new conceptual structures modelled on their forebears (see the works cited in chap. 1, n. 30). And much of what follows takes the form of a decomposition of the notion of modelling into more primi-

2. Lynch's exegesis of the quotation from Wittgenstein is this: "If the 'use' is the 'life' of an expression, it is not as though a meaning is 'attached' to an otherwise lifeless sign. We first encounter the sign in use or against the backdrop of a practice in which it has a use. It is already a meaningful part of the practice, even if the individual needs to learn the rule together with the other aspects of the practice. It is misleading to ask 'how we attach meaning' to the sign, since the question implies that each of us separately accomplishes what is already established by the sign's use in the language game. This way of setting up the problem is like violently wresting a cell from a living body and then inspecting the cell to see how life would have been attached to it" (1992b, 289). The most detailed study of conceptual practices (in mathematics) is Livingston 1986; see also Lynch's discussion (1992a, 243–47) of both that work and Bloor's critique of it.

3. The notion of discipline as a performative agent might seem odd to those accustomed to thinking of discipline as a constraint upon human agency, but I want (like Foucault) to recognize that discipline is productive. There could be no conceptual practice without the kind of discipline at issue; there could be only marks on paper.

tive elements. As it appears in my example, at least, it is useful to distinguish three stages within any given modelling sequence, which I will describe briefly in order to sketch out the overall form of my analysis. I argued in the preceding chapters that modelling has to be understood as an open-ended process, having in advance no determinate destination, and this is certainly true of conceptual practice. Part of modelling is thus what I call *bridging,* or the construction of a bridgehead, that tentatively fixes a vector of cultural extension to be explored. Bridging, however, is not sufficient to efface the openness of modelling. It is not enough in itself to define a new conceptual system on the basis of an old one, and it instead marks out a space for *transcription*—the copying of established moves from the old system into the new space fixed by the bridgehead (hence my use of the word "bridgehead"). And, if my example is a reliable guide, even transcription can be insufficient to complete the modelling process. What remains is *filling,* completing the new system in the absence of any clear guidance from the base model.

Now, this decomposition of modelling into bridging, transcription, and filling is at the heart of my analysis of conceptual practice, and I will be able to clarify what these terms mean when we come to my example. For the moment, though, I want to make a general remark about how they connect to issues of agency. As I conceive them, bridging and filling are activities in which scientists display choice and discretion, the classic attributes of human agency. Scientists are active in these phases of the modelling process, in Fleck's sense. Bridging and filling are *free moves,* as I shall say. In contrast, transcription is where discipline asserts itself, where the disciplinary agency just discussed carries scientists along, where scientists become passive in the face of their training and established procedures. Transcriptions, in this sense, are disciplined *forced moves.* Conceptual practice therefore has, in fact, the familiar form of a *dance of agency,* in which the partners are alternately the classic human agent and disciplinary agency.[4] And two points are worth

4. It might be useful to connect this remark to the earlier analysis of conceptual practice given by Pickering and Stephanides 1992 and the trajectory of emergence of the present book. As I mentioned in chapter 3, note 4, working through the Genoa quark-search experiments persuaded me that I needed to include material agency in my analysis, which led, in turn, to the concept of the dance of material and human agency discussed in the previous chapters. At the same time, I realized that the discussion of the intertwined active and passive aspects of conceptual practice given by Pickering and Stephanides 1992 was precisely an analysis of such a dance, if only one symmetrized the formulation by seeing human passivity there as the manifestation of disciplinary agency.

emphasizing here. First, this dance of agency, which manifests itself at the human end in the intertwining of free and forced moves in practice, is not optional. Practice has to take this form. The *point* of bridging as a free move is to invoke the forced moves that follow from it. Without such invocation, conceptual practice would be empty.[5] Second, the intertwining of free and forced moves implies what Gingras and Schweber (1986, 380) refer to (rather misleadingly) as a certain "rigidity" of conceptual "networks." I take this reference as a gesture toward the fact that scientists are not fully in control of where passages of conceptual practice will lead. Conceptual structures, one can say, relate to disciplinary agency much as do machines to material agency. Once one begins to tinker with the former, just as with the latter, one has to find out in practice how the resulting conceptual machinery will perform. It is precisely in this respect that dialectics of resistance and accommodation can arise in conceptual practice. To see how, though, requires some further discussion.

The constitutive role of disciplinary agency in conceptual practice is enough to guarantee that its end points are temporally emergent. One simply has to play through the moves that follow from the construction of specific bridgeheads and see where they lead. But this is not enough to explain the emergence of resistance, to get at how the workings of the mind lead the mind itself into problems. To get at this, one needs to understand what conceptual practice is for. I do not suppose that any short general answer to this question exists, but all of the examples that I can think of lead back to themes developed in the previous chapters, the themes of cultural multiplicity and the making and breaking of associations between diverse cultural elements. Let me give just two examples to illustrate what I have in mind.

In science, one prominent object of conceptual practice is bringing theoretical ideas to bear upon empirical data, to understand or explain the latter, to extract supposedly more fundamental information from them, or whatever. In *Constructing Quarks,* I argued that this process was indeed one of modelling, and now I would add four remarks. First, this process points once more to the multiplicity (and heterogeneity) of scientific culture. Data and theory have no necessary connection to one another; such connections as exist between them have to be made. Hence my second point: conceptual practice aims at making associa-

5. With a little work, one could connect this remark and the preceding argument to my earlier analyses (1981c, 1984b) of the ways in which expertise structures practice.

tions (translations, alignments) between such diverse elements—here data and theory, in a process that we have, in fact, already explored to a certain extent in relation to the Genoa experiments. Third, just because of the presence of the disciplinary partner in the dance of agency in conceptual practice, resistances can arise in the making of such associations. Because the destinations of conceptual practice cannot be known in advance, the pieces do not necessarily fit together as intended, as Morpurgo found in his quark searches. And fourth, as we saw in chapter 3, these resistances precipitate dialectics of resistance and accommodation, manglings that can bear upon conceptual structures as well as the form and performance of material apparatus.[6]

Of course, in chapter 3 I did not document and analyze how disciplines structure practice in theoretical science—I have just simply asserted it to be the case—and I do not propose to do so now. The reason for my reticence is that the disciplines at stake in all of the interesting cases that I know about—largely in recent theoretical elementary-particle physics—are sufficiently esoteric to make analysis and exposition forbiddingly daunting. I therefore propose instead to concentrate in this chapter on mathematics, and in particular on an example from the history of mathematics that is at once intellectually and historically interesting and simple, in that it draws only upon relatively low-level and already familiar disciplines in basic algebra and geometry. I hope thus to find an example of the mangle in action in conceptual practice that is accessible while being rich enough to point to further extensions of the analysis, in science proper as well as in mathematics. I will come to the example in a moment, but first some remarks are needed on mathematics in general.

Physics might be said to seek, among other things, somehow to describe the world; but what is mathematics for? Once more, I suppose that there is no general answer to this question, but I think that Latour

6. A point of clarification in relation to my earlier writings might be useful here. In Pickering 1989c, I discussed the open-ended extension of scientific culture in terms of a metaphor of "plasticity." I said that cultural elements were plastic resources for practice. The problem with this metaphor is that, if taken too seriously, it makes scientific practice sound too easy—one just keeps molding the bits of putty until they fit together. The upshot of the present discussion of disciplinary agency is that, unlike putty, pieces of conceptual culture keep transforming themselves in unpredictable ways after one has squeezed them (and evidently the same can be said of machinic culture: one can tinker with the material configuration of an apparatus, but that does not determine how it will turn out to perform). This is why achieving associations in practice is really difficult (and chancy).

(1987) makes some important and insightful moves. In his discussion of mathematical formalisms, Latour continually invokes metaphors of joining, linking, association, and alignment, comparing mathematical structures to railway turntables (239), crossroads (241), cloverleaf junctions (241), and telephone exchanges (242). His idea is, then, that such structures themselves serve as multipurpose translation devices, making connections between diverse cultural elements. And, as we shall see, this turns out to be the case in our example. The details are in sections 4.2 and 4.3, but the general point can be made in advance. If cultural extension in conceptual practice is not fully under the control of active human agents, due to the constitutive role of disciplinary agency, then the making of new associations—the construction of new telephone exchanges linking new kinds of subscribers—is nontrivial. Novel conceptual structures need to be tuned if they are to stand a chance of performing cooperatively in fields of disciplinary agency; one has to expect that resistances will arise in the construction of new conceptual associations, precipitating continuing dialectics of resistance and accommodation, manglings of modelling vectors—of bridgeheads and fillings, and even of the descriptions under which transcriptions are carried through, that is, of disciplines themselves.[7]

This is the process that we can now explore in an example taken from the history of mathematics. In the next two sections, we will be concerned with the work of the great Irish mathematician, Sir William Rowan Hamilton, and in particular with a brief passage of his mathematical practice that culminated on 16 October 1843 in the construction of his new mathematical system of quaternions. Before we turn to the study, however, I want to remark on the selection of this example.

7. I should mention one important aspect of mathematics that distinguishes it from science and to which I cannot pay detailed attention in this chapter, namely, mathematical *proof*. Here Lakatos's account (1976) of the development of Euler's theorem points once more to the mangle in conceptual practice. The exhibition of novel counterexamples to specific proofs of the theorem counts, in my terminology, as the emergence of resistances, and Lakatos describes very nicely the revision of proof procedures as open-ended accommodation to such resistances, with interactive stabilization amounting to the reconciliation of such procedures to given counterexamples. Other work in the history and philosophy of mathematics that points toward an understanding of practice as the mutual adjustment of cultural elements includes that of Philip Kitcher (1983, 1988), who argues that every mathematical practice has five components (1988, 299); Michael Crowe (1988, 1990; see n. 17 below); and Gaston Bachelard (Tiles 1984), who understands conceptual practice in terms of "resistances" (his word) and "interferences" between disjoint domains of mathematics.

As indicated above, it recommends itself on several accounts. The disciplinary agency manifest in Hamilton's work has a simple and familiar structure, which makes his work much easier to follow than that of present-day mathematicians or scientists. At the same time, Hamilton's achievement in constructing quaternions is of considerable historical interest. It marked an important turning point in the development of mathematics, involving as it did the first introduction of noncommuting quantities into the subject matter of the field, as well as the introduction of an exemplary set of new entities and operations, the quaternion system, that mutated over time into the vector analysis central to modern physics. Further, detailed documentation of Hamilton's practice is available.[8] Hamilton himself left several accounts of the passage of practice that led him to quaternions, especially a notebook entry written on the day of the discovery and a letter to John T. Graves dated the following day (Hamilton 1843a, 1843b, denoted by LTG and NBE hereafter; citations to page numbers are from the reprints of these in Hamilton 1967).[9] As Hamilton's biographer puts it, "These documents make the moment of truth on Dublin bridge [where Hamilton first conceived of the quaternion system] one of the best-documented discoveries in the history of mathematics" (Hankins 1980, 295). On this last point, some discussion is needed.

Hamilton's discovery of quaternions is not just well-documented, it is also much written about. Most accounts of Hamilton's algebraic researches contain some treatment of quaternions, and at least five accounts in the secondary literature rehearse to various ends Hamilton's own accounts more or less in their entirety (Hankins 1980, 295–300; O'Neill 1986, 365–68; Pycior 1976, chap. 7; van der Warden 1976; Whittaker 1945). I should therefore make it clear that what differentiates my account from others is that, as already indicated, I want to show that Hamilton's work can indeed be grasped within the more general understanding of agency and practice that I call the mangle. Together

8. I would have been entirely unaware of this, were it not for the work of Adam Stephanides, then a graduate student nominally under my supervision. Stephanides brought Hamilton's work to my attention by writing a very insightful essay emphasizing the open-endedness of Hamilton's mathematical practice, an essay that eventually turned into Pickering and Stephanides 1992.

9. Thus my primary source of documentation is a first-person narrative written after the event. This has to be understood as an edited rather than a complete account (whatever the latter might mean), but it is sufficient to exemplify the operation of the mangle in conceptual practice, which is my central concern.

with the discussion of forced moves and disciplinary agency, the open-endedness of modelling is especially important here, and in the narrative that follows, I seek to locate free moves in Hamilton's eventual route to quaternions by setting that trajectory in relation to his earlier attempts to construct systems of "triplets."

Section 4.2 describes the technical background to Hamilton's work. Section 4.3 analyzes his practice leading up to quaternions. Section 4.4 generalizes from the example in a discussion of temporal emergence and posthumanist decentering in conceptual practice, and of the mangling and interactive stabilization of conceptual structures, disciplines, and intentions. Since chapter 5 shifts the site of examination from science and mathematics to the factory, section 4.5 offers a quick overview of what has emerged about science in this and chapters 2 and 3. Finally, section 4.6 is a postscript that reviews David Bloor's account of Hamilton's metaphysics as an instance of SSK in its empirical application. Bloor argues that Hamilton's metaphysics was fixed by his vision of society; I argue instead that metaphysics, like everything else, is subject to mangling in practice.

4.2 FROM COMPLEX NUMBERS TO TRIPLETS

The early nineteenth century was a time of crisis in the foundations of algebra, centering on the question of how the "absurd" quantities—negative numbers and their square roots—should be understood (Hankins 1980, 248; Pycior 1976, chap. 4). Various moves were made in the debate over the absurd quantities, only one of which bears upon our story, and which serves to introduce the themes of cultural multiplicity and association as they will figure there. This was the move to construct an association between algebra and an otherwise disparate branch of mathematics, geometry, where the association in question consisted in establishing a *one-to-one correspondence* between the elements and operations of complex algebra and a particular geometrical system (Crowe 1985, 5–11). I need to go into some detail about the substance of this association, since it figured importantly in Hamilton's construction of quaternions.

The standard algebraic notation for a complex number is $x + iy$, where x and y are real numbers and $i^2 = -1$. Positive real numbers can be thought of as representing measurable quantities or magnitudes—a number of apples, the length of a rod—and the foundational problem in algebra was to think what -1 and i (and multiples thereof) might

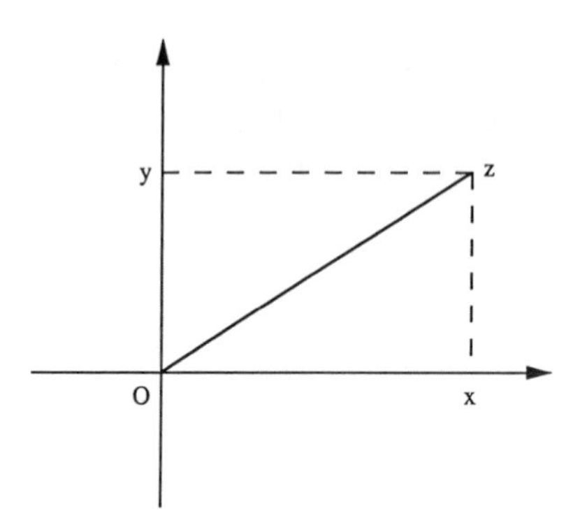

Fig. 4.1. Geometrical representation of the complex number $z = x + iy$ in the complex plane. The projections onto the x- and y-axes of the endpoint of the line Oz measure the real and imaginary parts of z, respectively.

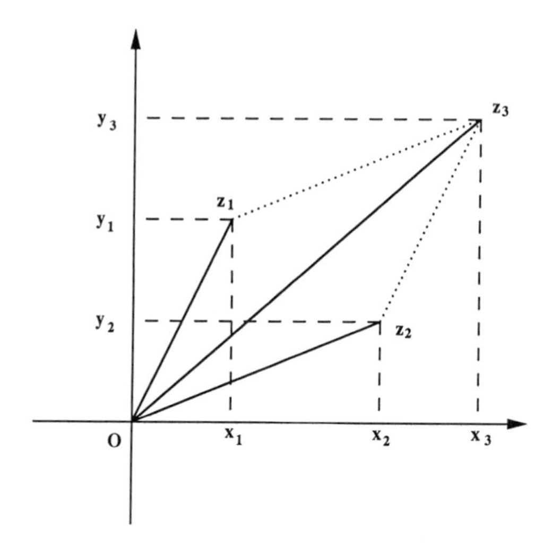

Fig. 4.2. Addition of complex numbers in the geometrical representation $z_3 = z_1 + z_2$. By construction, the real part of z_3 is the sum of the real parts of z_1 and z_2 ($x_3 = x_1 + x_2$), and likewise the imaginary part ($y_3 = y_1 + y_2$).

stand for. What sense can one make of $\sqrt{-1}$ apples? How many apples is that? The geometrical response to such questions was to think of x and y not as quantities or magnitudes, but as coordinates of the endpoint of a line segment terminating at the origin in some "complex" two-dimensional plane. Thus the x-axis of the plane measured the real component of a given complex number represented as such a line segment, and the y-axis the imaginary part, the part multiplied by i in the algebraic expression (fig. 4.1). In this way the entities of complex algebra were set in a one-to-one correspondence with geometrical line segments. Further, it was possible to put the operations of complex algebra in a similar relation with suitably defined operations upon line segments. Addition of line segments was readily defined on this criterion. In algebraic notation, addition of two complex numbers was defined as

$$(a + ib) + (c + id) = (a + c) + i(b + d),$$

and the corresponding rule for line segments was that the x-coordinate of the sum should be the sum of the x-coordinates of the segments to be summed, and likewise for the y-coordinate (fig. 4.2). The rule for subtraction could be obtained directly from the rule for addition—coordinates of line segments were to be subtracted instead of summed.

The rules for multiplication and division in the geometrical representation were more complicated, and we need only discuss that for multiplication, since this was the operation that became central in Hamilton's development of quaternions. The rule for algebraic multiplication of two complex numbers,

$$(a + ib)(c + id) = (ac - bd) + i(ad + bc),$$

followed from the usual rules of algebra, coupled with the peculiar definition of $i^2 = -1$. The problem was then to think what the equivalent might be in the geometrical representation. It proved to be statable as the conjunction of two rules. The product of two line segments is another line segment that (1) has a length given by the product of the lengths of the two segments to be multiplied, and that (2) makes an angle with the x-axis equal to the sum of the angles made by the two segments (fig. 4.3). From this definition, it is easy to check that multiplication of line segments in the geometrical representation leads to a result equivalent to the multiplication of the corresponding complex numbers in the algebraic representation.[10] Coupled with a suitably contrived

10. The easiest way to grasp these rules is as follows. In algebraic notation, any complex number $z = x + iy$ can be reexpressed as $r(\cos\theta + i\sin\theta)$, which can in turn be reinterpreted geometrically as a line segment of length r, subtending an angle θ with the x-axis at the origin. The product of two complex numbers z_1 and z_2 is therefore $r_1r_2(\cos\theta_1 + i\sin\theta_1)(\cos\theta_2 + i\sin\theta_2)$. When the terms in brackets are multiplied out and rearranged using standard trigonometric relationships, one arrives at $z_1z_2 = r_1r_2[\cos(\theta_1 + \theta_2) + i\sin(\theta_1 + \theta_2)]$, which can itself be reinterpreted geometrically as a line segment having a length that is the product of the lengths of the lines to be multiplied (part 1 of

definition of division in the geometrical representation, then, an association of one-to-one correspondence was achieved between the entities and operations of complex algebra and their geometrical representation in terms of line segments in the complex plane.

At least three important consequences for nineteenth-century mathematics flowed from this association. First, it could be said (though it could also be disputed) that the association solved the foundational problems centered on the absurd numbers. Instead of trying to understand negative and imaginary numbers as somehow measures of quantities or magnitudes of real objects, one should think of them geometrically, in terms of the orientation of line segments. A negative number, for example, should be understood as referring to a line segment lying along the negative (rather than positive) x-axis, a pure imaginary number as lying along the y-axis, and so on (fig. 4.4). Thus one could appeal for an understanding of the absurd numbers to an intuition of the possible differences in length and orientation of rigid bodies—sticks, say—in any given plane, and hence the foundational problem could be shown to be imaginary rather than real (so to speak).

Second, more practically, the geometrical representation of complex algebra functioned as a switchyard. Algebraic problems could be reformulated as geometrical ones, and thus perhaps solved, and vice versa. The third consequence of this association of algebra with geometry was that the latter, more clearly than the former, invited extension. Complex algebra was a self-contained field of mathematical practice; geometry, in contrast, was by no means confined to the plane. The invitation, then, was to extend the geometrical representation of complex-number theory from a two- to a three-dimensional space, and to somehow carry along a three-place algebraic equivalent with it, maintaining the association already constructed in two dimensions. On the one hand, this extension could be attempted in a spirit of play, just to see what could be achieved. On the other, there was a promise of utility. The hope was to construct an algebraic replica of transformations of line segments in three-dimensional space, and thus to develop a new and possibly useful algebraic system appropriate to calculations in three-dimensional geometry, "to connect, in some new and useful (or at least interesting) way *calculation* with *geometry,* through some *extension* [of the association achieved

---

the rule) and making an angle with the x-axis equal to the sum of angles made by the lines to be multiplied (part 2).

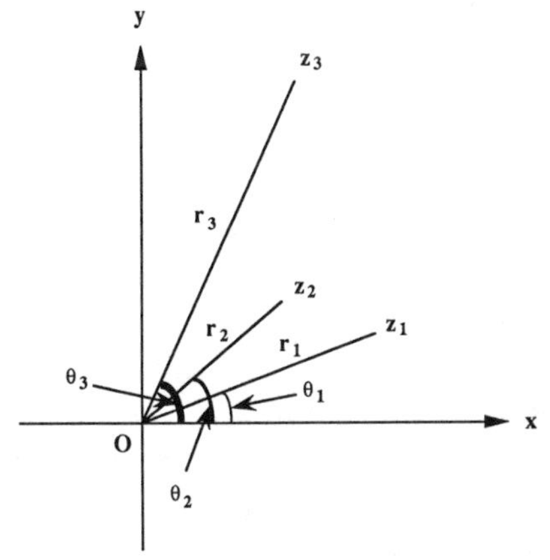

Fig. 4.3. Multiplication of complex numbers in the geometrical representation $z_3 = z_1 \times z_2$. Here the lengths of line segments are multiplied ($r_3 = r_1 \times r_2$), while the angles subtended with the x-axis by line segments are added ($\theta_3 = \theta_1 + \theta_2$).

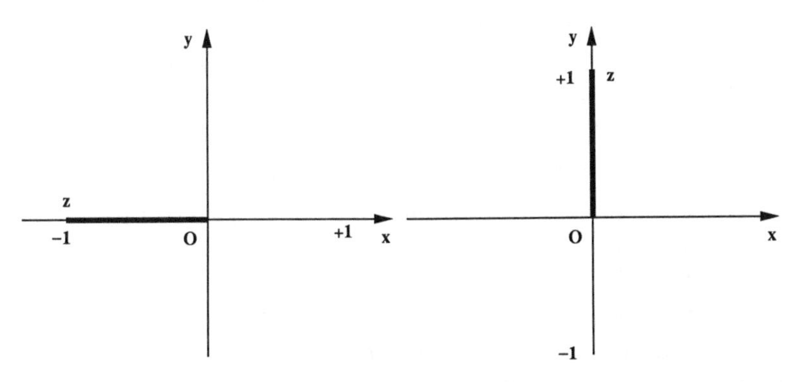

Fig. 4.4. Geometrical representations of absurd numbers: (left) $z = -1$; (right) $z = i = \sqrt{-1}$.

in two dimensions], to *space of three dimensions,*" as Hamilton put it (1853, 135).[11]

Hamilton was involved in the development of complex algebra from the late 1820s onward. He worked both on the foundational problems just discussed (developing his own approach to them via his "Science of

11. The perceived need for an algebraic system that could represent elements and operations in three-dimensional space more perspicuously than existing systems is discussed

Pure Time" and a system of "couples" rather than through geometry; I return to this topic in section 4.6) and on the extension of complex numbers from two- to three-place systems, or "triplets," as he called them. His attempts to construct triplet systems in the 1830s were many and various, but Hamilton regarded them all as failures (Hamilton 1967, 3–100, 117–42; Hankins 1980, 245–301; Pycior 1976, chaps. 3–6). In 1843, after a period of work on other topics, he returned to the challenge once more. Again he failed to achieve his goal, but this time he did not come away empty-handed. Instead of constructing a three-place or three-dimensional system, he quickly arrived at the four-place quaternion system that he regarded as his greatest mathematical achievement and to which he devoted the remainder of his life's work. This is the passage of practice that I want to analyze in detail.

4.3 CONSTRUCTING QUATERNIONS

On 16 October 1843, Hamilton set down in a notebook his recollection of his path to quaternions. The entry begins: "I, this morning, was led to what seems to me a theory of *quaternions,* which may have interesting developments. *Couples* being supposed known, and known to be representable by points in a plane, so that $\sqrt{-1}$ is perpendicular to 1, it is natural to conceive that there may be another sort of $\sqrt{-1}$, perpendicular to the plane itself. Let this new imaginary be j; so that $j^2 = -1$, as well as $i^2 = -1$. A point x, y, z in space may suggest the triplet $x + iy + jz$" (NBE, 103).

I can begin my commentary on this passage by noting that a process of modelling was, as usual, constitutive of Hamilton's practice. As is evident from these opening sentences, he did not attempt to construct a three-place mathematical system out of nothing. Instead he sought to move into the unknown from the known, to find a creative extension of the two-place systems already in existence. Further, as will become evident as we go along, the process of cultural extension through modelling

in Crowe 1985, 3–12. Though Hamilton wrote of his desire to connect calculation with geometry some years after the event, he recalled in the same passage that he was encouraged to persevere in the face of difficulties by his friend John T. Graves, "who felt the wish, and formed the project, to surmount them in some way, as early, or perhaps earlier than myself" (1853, 137; page citations to this essay refer to the reprint in Hamilton 1967). Hamilton's common interest with Graves in algebra dated back to the late 1820s (Hankins 1980, chap. 17), so there is no reason to doubt that this utilitarian interest did play a role in Hamilton's practice. See also O'Neill 1986.

was, in this instance as in general, an open-ended one: in his work on triplet systems that culminated in the construction of quaternions, Hamilton tried out a large number of different extensions of complex algebra and geometry. Now I need to talk about how Hamilton moved around in this open-ended space, a discussion that will lead us into the tripartite decomposition of modelling mentioned in section 4.1.

In his reference above to "points in a plane," Hamilton first invokes the geometrical representation of complex algebra, and the extension that he considers is to move from thinking about line segments in a plane to thinking about line segments in a three-dimensional space. I say that, in so doing, he established a *bridgehead* to a possible three-dimensional extension of complex algebra. As already stated and as discussed further below, the significance of such a bridging operation is that it marks a particular destination for modelling; at the moment, I want to emphasize two points that I suspect are general about bridging. First, however natural Hamilton's move from the plane to three-dimensional space might seem, it is important to recognize that it was by no means forced upon him. In fact, in his earlier attempts at triplet systems, he had proceeded differently, often working first in terms of an algebraic model and only toward the end of his calculations attempting to find geometrical representations of his findings, representations that were quite dissimilar from that with which he begins here (Hamilton 1853, 126–32).[12] In this sense, the act of fixing a bridgehead is an active or free move that serves to cut down the indefinite openness of modelling. My second point follows from this. Such free moves need to be seen as tentative and revisable trials that carry with them no guarantee of success. Just as Hamilton's earlier choices of bridgeheads had, in his own estimation, led to failure, so might this one. His only way of assessing this particular choice was to work with it and on it—to see what he could make of it. Similar comments apply to the second model that structured Hamilton's practice. This was the standard algebraic formulation of complex numbers, which he extends in the above quotation to a three-place system by mov-

12. In such attempts, the intention to preserve any useful association of algebra and geometry does not seem to be central: Hamilton's principal intent was simply to model the development of a three-place algebraic system on his existing two-place system of couples. Because the construction of associations in a multiple field plays a key role in my analysis, I should note that attention to this concept illuminates even these principally algebraic attempts. Hamilton found it necessary to transcribe parts of his development of couples *piecemeal,* and the goal of reassembling (associating) the disparate parts of the system that resulted again led to the emergence of resistance.

ing from the usual $x + iy$ notation to $x + iy + jz$. This seems like another natural move to make. But again, when set against Hamilton's earlier work on triplets, it is better seen as the establishment of a bridgehead in a tentative free move.[13]

One more remark before returning to Hamilton's recollections. I noted above that complex algebra and its geometrical representation were associated with one another in a relation of one-to-one correspondence, and an intent to preserve that association characterized the passage of Hamilton's practice presently under discussion. In the quotation, he sets up a one-to-one correspondence between the *elements* defined in his two bridging moves—between the algebraic notation $x + iy + jz$ and suitably defined three-dimensional line segments. In the passage that follows, he considers the possibility of preserving the same association of mathematical *operations* in the two systems. This is where the analysis of modelling becomes interesting, where disciplinary agency comes into play and the possibility of resistance in conceptual practice thus becomes manifest. Hamilton's notebook entry continues:

> The square of this triplet [$x + iy + jz$] is on the one hand $x^2 - y^2 - z^2 + 2ixy + 2jxz + 2ijyz$; such at least it seemed to me at first, because I assumed $ij = ji$. On the other hand, if this is to represent the third proportional to 1, 0, 0 and x, y, z, considered as *indicators of lines,* (namely the lines which end in the points having these coordinates, while they begin at the origin) and if this third proportional be supposed to have its length a third proportional to 1 and $\sqrt{(x^2 + y^2 + z^2)}$, and its distance twice as far angularly removed from 1, 0, 0 as x, y, z; then its real part ought to be $x^2 - y^2 - z^2$ and its two imaginary parts ought to have for coefficients 2xy and 2xz; thus the term 2ijyz appeared de trop, and I was led to assume at first $ij = 0$. However I saw that this difficulty would be removed by supposing that $ji = -ij$. (NBE, 103)

This passage requires some exegesis. Here Hamilton begins to think about mathematical operations on the three-place elements that his bridgeheads have defined, and in particular about the operation of multiplication, specialized initially to that of squaring an arbitrary triplet.

13. The foundational significance of Hamilton's couples was precisely that the symbol *i* did not appear in them, and was therefore absent from the attempts at triplets discussed in note 12 above. A typical bridging move there was to go from couples written as (*a, b*) to triplets written (*a, b, c*).

He works first in the purely algebraic representation, and if, for clarity, we write $t = x + iy + jz$, he finds

$$t^2 = x^2 - y^2 - z^2 + 2ixy + 2jxz + 2ijyz. \qquad (1)$$

This equation follows automatically from the laws of standard algebra, coupled with the usual definition that $i^2 = -1$ and the new definition $j^2 = -1$ that was part of Hamilton's algebraic bridgehead. In this instance, then, we see that the primitive notion of modelling can be partly decomposed into two more transparent operations, bridging and *transcription,* where the latter amounts to the copying of an operation defined in the base model—in this instance the rules of algebraic multiplication—into the system set up by the bridgehead. And this, indeed, is why I use the word "bridgehead": it defines a point to which attributes of the base model can be transferred, a destination for modelling, as I put it earlier. We can note here that, just as it is appropriate to think of fixing a bridgehead as an active, free move, it is likewise appropriate to think of transcription as a sequence of passive, forced moves, a sequence of moves—resulting here in equation 1—that follow from what is already established concerning the base model. And we can note further that the surrender of agency on Hamilton's part is equivalent to the assumption of agency by discipline. While Hamilton was indeed the person who thought through and wrote out the multiplications in question, he was not free to choose how to perform them. Anyone already disciplined in algebraic practice, then or now, can check that Hamilton (and I) have done the multiplication correctly. This then is our first example of the dance of agency in conceptual practice, in which disciplinary agency carried Hamilton (and carries us) beyond the fixing of a bridgehead.

The disciplined nature of transcription is what makes possible the emergence of resistance in conceptual practice, but before we come to that, we should note that the decomposition of modelling into bridging and transcription is only partial. Equation 1 still contains an undefined quantity—the product ij—that appears in the last term of the right-hand side. This was determined neither in Hamilton's first free move nor in the forced moves that followed. The emergence of such "gaps" is, I believe, another general feature of the modelling process: disciplinary agency is insufficient to carry through the processes of cultural extension that begin with bridging. Gaps appear throughout Hamilton's work on triplets, for example, and one of his typical responses was that which

I call *filling,* meaning the assignment of values to undefined terms in further free moves.[14] Resuming the initiative in the dance of agency, here Hamilton could have, say, simply assigned a value to the product *ij* and explored where that led him through further forced moves. In this instance, though, he proceeded differently.

The sentences that begin "On the other hand, if this is to represent the third proportional . . ." refer to the operation of squaring a triplet in the geometrical rather than in the algebraic representation. Considering a triplet as a line segment in space, Hamilton was almost in a position to transcribe onto his new bridgehead the rules for complex multiplication summarized in section 4.2, but, although not made explicit in the passage, one problem remained. While the first rule concerning the length of the product of lines remained unambiguous in three-dimensional space, the second, concerning the orientation of the product line, did not. Taken literally, it implied that the angle made by the square of any triplet with the x-axis was twice the angle made by the triplet itself—"twice as far angularly removed from 1, 0, 0 as x, y, z"—but it in no way specified the orientation of the product line in space. Here disciplinary agency again left Hamilton in the lurch. Another gap thus arose in moving from two to three dimensions, and, in this instance, Hamilton responded with a characteristic, if unacknowledged, filling move.

He further specified the rule for multiplication of line segments in space by enforcing the new requirement that the square of a triplet remain in the plane defined by itself and the x-axis (this is the only way in which one can obtain his stated result for the square of a triplet in the geometrical representation). As usual, this move seems natural enough, but the sense of naturalness is easily shaken when taken in the context of Hamilton's prior practice. One of Hamilton's earliest attempts at triplets, for example, represented them as lines in three-dimensional space, but multiplication was defined differently in that attempt.[15] Be that as it may, this particular filling move sufficed and was designed to make possible a series of forced transcriptions from the two- to the three-dimensional versions of complex algebra that enabled Hamilton to com-

14. See, for example, his development of rules for the multiplication of couples (Hamilton 1837, 80–83; page citations refer to the 1967 reprint).

15. Hamilton (1853, 139–40) cites his notes of 1830 as containing an attempt at constructing a geometrical system of triplets by denoting the end of a line segment in spherical polar coordinates as $x = r\cos\theta$, $y = r\sin\theta\cos\phi$, $z = r\sin\theta\sin\phi$, and extending the rule of multiplication from two to three dimensions as $r'' = rr'$, $\theta'' = \theta + \theta'$, $\phi'' = \phi + \phi'$. This addition rule for the angle ϕ breaks the coplanarity requirement at issue.

pute the square of an arbitrary triplet. Surrendering once more to the flow of discipline, he found that the "real part [of the corresponding line segment] ought to be $x^2 - y^2 - z^2$ and its two imaginary parts ought to have for coefficients 2xy and 2xz." Or, returning this result to purely algebraic notation,[16]

$$t^2 = x^2 - y^2 - z^2 + 2ixy + 2jxz. \qquad (2)$$

Now, there is a simple difference between equations 1 and 2, both of which represent the square of a triplet, but calculated in different ways. The two equations are identical except that the problematic term $2ijyz$ of equation 1 is absent from equation 2. This, of course, is just the kind of thing that Hamilton was looking for to help him in defining the product ij, and we will examine the use he made of it in a moment. First, it is time to talk about *resistance*. The two base models that Hamilton took as his points of departure—the algebraic and geometrical representations of complex numbers—were associated in a one-to-one correspondence of elements and operations. Here, however, we see that, as so far extended by Hamilton, the three-place systems had lost this association. The definition of a square in the algebraic system (equation 1) differed from that computed via the geometrical representation (equation 2). The association of "calculation with geometry" that Hamilton wanted to preserve had been broken; a resistance to the achievement of his goal had appeared. And, as I have already suggested, the precondition for the emergence of this resistance was the constitutive role of disciplinary agency in conceptual practice and the consequent intertwining of free and forced moves in the modelling process. Hamilton's free moves had determined the directions that his extensions of algebra and geometry would take in the indefinitely open space of modelling, but the forced moves intertwined with them had carried those extensions along to the point at which they collided in equations 1 and 2. This, I think, is how "the workings of the mind lead the mind itself into problems." We can now move from resistance itself to a consideration of the dialectic of resistance and accommodation in conceptual practice—in other words, to the mangle.

16. One route to this result is to write the triplet t in spherical polar notation. According to the rule just stated, on squaring, the length of the line segment goes from r to r^2, the angle θ doubles, while the angle ϕ remains the same. Using standard trigonometric relations to express $\cos 2\theta$ and $\sin 2\theta$ in terms of $\cos\theta$ and $\sin\theta$, one can then return to x, y, z notation and arrive at equation 2.

The resistance that Hamilton encountered in the disparity between equations 1 and 2 can be thought of as an instance of a generalized version of the Duhem problem.[17] Something had gone wrong somewhere in the process of cultural extension—the pieces did not fit together as desired—but Hamilton had no principled way of knowing where. What remained for him to do was to tinker with the various extensions in question—with the various free moves he had made, and thus with the sequences of forced moves that followed from them—in the hope of getting around the resistance that had arisen and achieving the desired association of algebra and geometry. He was left, as I say, to seek some accommodation to resistance. Two possible starts toward accommodation are indicated in the passage last quoted, both of which amounted to further fillings-in of Hamilton's extended algebraic system, and both of which led directly to an equivalence between equations 1 and 2. The most straightforward accommodation was to set the product ij equal to zero.[18] An alternative, less restrictive, but more dramatic and eventually more far-reaching move also struck Hamilton as possible. It was to abandon the assumption of commutation between i and the new square root of -1, j.[19] In ordinary algebra, this assumption—which is to say that $ab = ba$—was routine. Hamilton entertained the possibility, instead, that $ij = -ji$. This did not rule out the possibility that both ij

17. See chapter 3, note 19. The Duhem problem is usually formulated in terms of mismatches between data and theoretical predictions. As far as I am aware, the only prior discussion of it as it bears on purely mathematical-conceptual practice is to be found in Crowe 1990, 441–43, which argues, following Lakatos 1976, that contradictions between proofs and counterexamples need not necessarily disable the former. Crowe 1988 also discusses the extension of Duhem's ideas about physics to mathematics, and gestures repeatedly, though without detailed documentation or analysis, to the interactive stabilization of axioms and theorems proved within them, and of mathematical systems and the results to which they lead. I thank Professor Crowe for drawing my attention to these essays.

18. The following day, Hamilton described the idea of setting $ij = 0$ as "odd and uncomfortable" (LTG, 107). He offered no reasons for this description, and it is perhaps best understood as written from the perspective of his subsequent achievement. The quaternion system preserved the geometrical rule of multiplication that the length of the product was the product of the lengths of the lines multiplied. Since in the geometrical representation both i and j have unit length, the equation $ij = 0$ violates this rule. Here we have a possible example of the retrospective reconstruction of accounts in the rationalization of free moves.

19. Pycior (1976, 147) notes that Hamilton had been experimenting with noncommuting algebras as early as August 1842, though he then tried the relations $ij = j$, $ji = i$. Hankins (1980, 291–92) detects a possible influence of a meeting between Hamilton and the German mathematician Gotthold Eisenstein in the summer of 1843.

and *ji* were zero, but even without this being the case, it did guarantee that the problematic term $2ijyz$ of equation 1 vanished, and thus constituted a successful accommodation to the resistance that had emerged at this stage.[20]

Hamilton thus satisfied himself that he could maintain the association between his algebraic and geometrical three-place systems by the assumption that *i* and *j* did not commute, at least as far as the operation of squaring a triplet was concerned. His next move was to consider a less restrictive version of the general operation of multiplication, working through, as above, the operation of multiplying two coplanar but otherwise arbitrary triplets. Again, he found that the results of the calculation were the same in the algebraic and geometrical representations as long as he assumed either $ij = 0$ or $ij = -ji$ (NBE, 103). Hamilton then moved on to consider the fully general instance of multiplication in the new formalism, the multiplication of two arbitrary triplets (NBE, 103–4). As before, he began in the algebraic representation. Continuing to assume $ij = -ji$, he wrote

$$(\mathrm{a} + \mathrm{ib} + \mathrm{jc})(\mathrm{x} + \mathrm{iy} + \mathrm{jz}) = \mathrm{ax} - \mathrm{by} - \mathrm{cz} + \mathrm{i}(\mathrm{ay} + \mathrm{bx}) + \mathrm{j}(\mathrm{az} + \mathrm{cx}) + \mathrm{ij}(\mathrm{bz} - \mathrm{cy}). \quad (3)$$

He then turned back to thinking about multiplication within the geometrical representation, where a further problem arose. Recall that, in defining the operation of squaring a triplet, Hamilton had found it necessary to make a filling free move, assuming that the square lay in the plane of the original triplet and the x-axis. This filling move was sufficient to lead him through a series of forced moves to the calculation of the product of two arbitrary but coplanar triplets. But it was insufficient to define the orientation in space of the product of two completely arbitrary triplets: in general, one could not pass a plane through any two triplets and the x-axis. Once more, Hamilton could have attempted a filling move here, concocting some rule for the orientation of the product line in space, say, and continuing to apply the sum rule for the angle made by the product with the x-axis. In this instance, however, he followed a different strategy.

Instead of attempting the transcription of the two rules that fully specified multiplication in the standard geometrical representation of

20. If one multiplies out the terms of equation 1, paying attention to the order of factors, the coefficient of *yz* in the last term on the right-hand side becomes $(ij + ji)$; Hamilton's assumption makes this coefficient zero.

complex algebra, he began to work only in terms of the first rule—that the length of the product line segment should be the product of the lengths of the line segments to be multiplied. Transcribing this rule to three dimensions, and working for convenience with squares of lengths, or "square moduli," rather than lengths themselves, he could surrender his agency to Pythagoras's theorem and write the square modulus of the left-hand side of equation 3 as $(a^2 + b^2 + c^2)(x^2 + y^2 + z^2)$ (another forced move).[21] Now he had to compute the square of the length of the right-hand side. Here the obstacle to the application of Pythagoras's theorem was the quantity ij again appearing in the last term. If Hamilton assumed that $ij = 0$, the theorem could be straightforwardly applied, and gave a value for the square modulus of $(ax - by - cz)^2 + (ay + bx)^2 + (az + cx)^2$. The question now was whether these two expressions for the lengths of the line segments appearing on the two sides of equation 3 were equal. Hamilton multiplied them out and rearranged the expression for the square modulus of the left-hand side, and found that it in fact differed from that on the right-hand side by a factor of $(bz - cy)^2$. Once again a resistance had arisen, now in thinking about the product of two arbitrary triplets in, alternatively, the algebraic and geometrical representations. Once more, the two representations, extended from two- to three-place systems, led to different results. And once more, Hamilton looked for some accommodation to this resistance, for some way of making the two notions of multiplication equivalent, as they were in two dimensions.

The new resistance was conditional on the assumption that $ij = 0$. The question, then, was whether some other assignment of ij might succeed in balancing the moduli of the left- and right-hand sides of equation 3.[22] And here Hamilton made a key observation. The superfluous term in the square modulus of the left-hand side of equation 3, $(bz - cy)^2$, was the square of the coefficient of ij on the right-hand side. The two computations of the square modulus could thus be made to balance by assuming not that the product of i and j vanished, but that it was some third quantity k, a *"new imaginary"* (NBE, 104), different again

21. According to Pythagoras's theorem, the square modulus of a line segment is simply the sum of the squares of the coordinates of its endpoints, meaning the coefficients of 1, i, and j in algebraic notation.

22. Strictly speaking, this is too deterministic a formulation. The question really was whether any amount of tinkering with bridgeheads, fillings, and so on could get past this point without calling up this or another resistance.

from i and j, in such a way that Pythagoras's theorem could be applied to it, too.

The introduction of the new imaginary k, defined as the product of i and j, thus constituted a further accommodation by Hamilton to an emergent resistance in thinking about the product of two arbitrary triplets in terms of the algebraic and geometrical representations at once, and one aspect of this particular accommodation is worth emphasizing. It amounted to a drastic shift of bridgehead in both systems of representation (recall that I stressed the revisability of bridgeheads earlier). More precisely, it consisted in defining a new bridgehead leading from two-place representations of complex algebra to not three- but four-place systems—the systems that Hamilton quickly called *quaternions*. Thus, within the algebraic representation, the basic entities were extended from two to four, from 1, i to 1, i, j, k, while within the geometrical representation, as Hamilton wrote the next day, "there dawned on me the notion that we must admit, in some sense, a *fourth dimension* of space" (LTG, 108)—with the fourth dimension, of course, mapped by the new k-axis.

We will consider this shift in bridgehead further in the next section; for now, we can observe that Hamilton had still not completed the initial development of quaternions. The quantity k^2 remained undefined at this stage, as did the various products of i and j with k, excepting those intrinsic to his new bridgehead $ij = k$. Hamilton fixed the latter products by a combination of filling assumptions and forced moves following from relations already fixed.

> I saw that we had probably $ik = -j$, because $ik = iij$ and $i^2 = -1$; and that in like manner we might expect to find $kj = ijj = -i$; from which I thought it likely that $ki = j$, $jk = i$, because it seemed likely that if $ji = -ij$, we should have also $kj = -jk$, $ik = -ki$. And since the order of these imaginaries is not indifferent, we cannot infer that $k^2 = ijij$ is $+1$, because $i^2 \times j^2 = -1 \times -1 = +1$. It is more likely that $k^2 = ijij = -iijj = -1$. And in fact this last assumption is necessary, if we would conform the multiplication to the law of multiplication of moduli. (LTG, 108)

Hamilton then checked whether the algebraic version of quaternion multiplication under the above assumptions, including $k^2 = -1$, led to results in accordance with the rule of multiplication concerning products of lengths in the geometrical representation ("the law of multiplication of moduli"), and found that it did. Everything in his quaternion system was thus now defined in such a way that the laws of multiplication in both the algebraic and geometrical versions ran without resistance into one another. Through the move to four-place systems, Hamilton had finally found a successful accommodation to the resistances that had stood in the way of his three-place extensions. The outcome of this dialectic was the general rule for quaternion multiplication (LTG, 108):[23]

$$(a, b, c, d)(a', b', c', d') = (a'', b'', c'', d''),$$

where

$$\begin{aligned} a'' &= aa' - bb' - cc' - dd', \\ b'' &= ab' + ba' + cd' - dc', \\ c'' &= ac' + ca' + db' - bd', \\ d'' &= ad' + da' + bc' - cb'. \end{aligned}$$

With these algebraic equations, and the geometrical representation of them, Hamilton had, in a sense, achieved his goal of associating calculation with geometry, and I could therefore end my narrative here. But before doing so, I want to emphasize that the qualifier "in a sense" is significant. It marks the fact that what Hamilton had achieved was a *local* association of calculation with geometry rather than a global one. He had constructed a one-to-one correspondence between a particular algebraic system and a particular geometric system, not an all-purpose link between algebra and geometry considered as abstract, all-encompassing entities. And this remark makes clear the fact that one important aspect of Hamilton's achievement was to redefine, partially at least, the cultural space of future mathematical and scientific practice: more new associations remained to be made if quaternions were ever to be "delocalized" and linked into the overall flow of mathematical and scientific practice, requiring work that would, importantly, have been inconceivable in advance of Hamilton's construction of quaternions.

As it happens, from 1843 onward Hamilton devoted most of his productive energies to this task, and both quaternions and the principle of noncommutation that they enshrined were taken up progressively by many sections of the scientific and mathematical communities (Hankins

23. (a, b, c, d) was Hamilton's notation for an arbitrary quaternion. In the geometrical representation, the coordinates of the endpoint of a line segment in four-dimensional space are given here; in algebraic notation, this same quaternion would be written $a + ib + jc + kd$.

1980, chap. 23; Crowe 1985, chaps. 4–7). Here I will discuss one last aspect of Hamilton's practice that can serve to highlight the locality of the association embodied in quaternions. Earlier I described Hamilton's organizing aim as that of connecting calculation with geometry. And, as just discussed, quaternions did serve to bring algebraic calculation to *a* geometry—to the peculiar four-dimensional space mapped by 1, i, j, and k. Unfortunately this was not the geometry for which calculation was desired. The promise of triplet—not quaternion—systems had been that they would bring algebra to bear upon the real three-dimensional world of interest to mathematicians and physicists. In threading his way through the dialectic of resistance and accommodation, Hamilton had, in effect, left that world behind. Or, to put it another way, his practice had, as so far described, served to displace resistance rather than fully to accommodate to it. Technical resistances in the development of three-place mathematical systems had been transmuted into a resistance between moving from Hamilton's four-dimensional world to the three-dimensional world of interest. It was not evident how the two worlds might be related to one another. This was one of the first problems that Hamilton addressed once he had arrived at his algebraic formulation of quaternions.

In his letter to John T. Graves dated 17 October 1843, Hamilton outlined a new geometrical interpretation of quaternions that served to connect them back to the world of three dimensions. This new interpretation was a straightforward but consequential redescription of the earlier four-dimensional representation. Hamilton's idea was to think of an arbitrary quaternion (a, b, c, d) as the sum of two parts: a real part, a, which was a pure real number and had no geometrical representation, and an imaginary part, the triplet, $ib + jc + kd$, which was to be represented geometrically as a line segment in three-dimensional space.[24] Having made this split, Hamilton was then in a position to spell out rules for multiplication of the latter line segments, which he summarized as follows:

> Finally, we may always decompose the latter problem [the multiplication of two arbitrary triplets] into these two others; to multiply two pure imaginaries which agree in direction, and to multiply two which are at

24. The origin of this distinction between real and imaginary parts of quaternions lay in the differences between the equations just quoted defining quaternion multiplication: the equation defining a'' has three minus signs while those defining b'', c'', and d'' have two plus signs and only one minus.

> right angles with each other. In the first case, the product is a pure negative, equal to the product of the lengths or moduli with its sign changed. In the second case, the product is a pure imaginary of which the length is the product of the lengths of the factors, and which is perpendicular to both of them. The distinction between one such perpendicular and its opposite may be made by the rule of rotation [stated earlier in this letter].
>
> There seems to me to be something analogous to *polarized intensity* in the pure imaginary part; and to *unpolarized energy* (indifferent to direction) in the real part of a quaternion: and thus we have some slight glimpse of a future Calculus of Polarities. This is certainly very vague, but I hope that most of what I have said above is clear and mathematical. (LTG, 110)

These strange rules for the multiplication of three-dimensional line segments—in which the product of two lines might be, depending on their relative orientation, a number or another line or some combination of the two—served to align quaternions with mathematical and scientific practice concerned with the three-dimensional world.[25] Nevertheless, the association of algebra with geometry remained local. No contemporary physical theories, for example, spoke of entities in three-dimensional space obeying Hamilton's rules. It therefore still remained to find out in practice whether quaternions could be delocalized to the point at which they might become useful. With hindsight, one can pick out from the rules of multiplication a foreshadowing of modern vector analysis with its "dot" and "cross" products, and in the references to "polarized intensity" and "unpolarized energy," one can find a gesture toward electromagnetic theory, where quaternions and vector analysis found their first important use. But, as Hamilton wrote, unlike the mathematics of quaternions this "slight glimpse of the future" was, in 1843, "certainly very vague." It was only in the 1880s, after Hamilton's death, that Josiah Willard Gibbs and Oliver Heaviside laid out the fundamentals of vector analysis, dismembering the quaternion system into more useful parts in the process (Crowe 1985, chap. 5). This key moment in the delocalization of quaternions was also the moment of their disintegration.

25. Note that this geometric interpretation included a handedness rule—a "rule of rotation"—which reversed the sign of the product of perpendicular lines when the order of their multiplication was reversed, thus explaining algebraic noncommutation in much the same way as the two-dimensional geometrical representation of complex numbers had explained the "absurd" negative and imaginary quantities.

4.4 CONCEPTS AND THE MANGLE

I have come to the end of my story of Hamilton and quaternions, and the analysis that I have interwoven with the narrative is complex enough, I think, to warrant a general summary and even a little further elaboration.

My overall object in this chapter has been to get to grips with the specifically conceptual aspects of scientific practice (I continue to use "scientific" as an umbrella term that includes mathematics). My point of departure has been the traditional one, an understanding of conceptual extension as a process of modelling. Thus I have tried to show that complex algebra and its geometrical representation in the complex plane were both constitutive models in Hamilton's practice. But I have gone beyond the tradition in two ways. First, instead of treating modelling (metaphor, analogy) as a primitive term, I have suggested that it bears further analysis and decomposition into the three phases of bridging, transcription, and filling. I have exemplified these phases and their interrelation in Hamilton's work, and I have tried to show how the openness of modelling is tentatively cut down by human discretionary choices (by human agency, traditionally conceived) in bridging and filling, and by disciplinary agency (disciplined, machinelike human agency) in transcription. I have also exemplified the fact that these two aspects of modelling—active and passive from the perspective of the human actor—are inextricably intertwined inasmuch as the object of constructing a bridgehead, for example, is, as I have stressed, to load onto it disciplined practices already established around the base model. Conceptual practice thus has the quality, familiar from earlier chapters, of a dance of agency, this time between the discretionary human agent and what I have been calling disciplinary agency. The constitutive part played by disciplinary agency in this dance guarantees that the free moves of human agents—bridging and filling—carry those agents along trajectories that cannot be foreseen in advance, that have to be found out in practice.

My second step beyond traditional conceptions of modelling has been to note that it does not proceed in a vacuum. Issues of cultural multiplicity surface again here. My suggestion is that conceptual practice is organized around the production of associations, the making of connections, and the creation of alignments between disparate cultural elements, where, in the present instance, the association in question was that between three-place algebras and three-dimensional geometries (initially, at least). And here the key observation is that the entanglement of disciplinary agency in practice makes the achievement of such associations nontrivial in the extreme. Hamilton wanted to extend algebra and geometry into three dimensions while maintaining a one-to-one correspondence of elements and operations between them, but neither he nor anyone subsequently has been able to do it. Resistance thus emerges in conceptual practice in relation to intended associations, and precipitates the dialectics of accommodation and further resistance that I call the mangle. Now I want to discuss just what gets mangled.

Most obviously mangled in Hamilton's practice were the modelling vectors that he pursued. In the face of resistance, he tinkered with choices of bridgeheads and fillings, tuning, one can say, the directions along which complex algebra and its geometrical representation were to be extended. And, as we saw, this mangling of modelling vectors eventually (not at all necessarily) met with success. In the quaternion system, he arrived at an association of one-to-one correspondence of elements and operations between an extension of complex algebra and an extension of its geometrical representation. And, taking over a phrase already exemplified in the preceding chapters, this achievement constituted an *interactive stabilization* of the specific free moves and the associated forced moves that led up to it. *This* particular bridgehead, coupled with *these* particular transcriptions and fillings, defined the vector along which complex algebra should be extended, and similarly for the associated geometry. Exactly how existing conceptual structures should be extended was, then, the upshot of the mangle—as was the precise structure of the quaternion system that these particular extensions defined.

Here it is worth pausing to reiterate for conceptual practice two points that I have already made in respect of captures and framings of material agency. First, the precise trajectory and endpoint of Hamilton's practice were in no way given in advance. Nothing prior to that practice determined its course. Hamilton had, in the real time of his mathematical work, to fix bridgeheads and fillings and to find out where they led via disciplined transcriptions. He had, further, to find out in real time just what resistances would emerge relative to intended conceptual alignments—such resistances again could not be foreseen in advance—and to make whatever accommodations he could find to them, with the success or failure of such accommodations itself only becoming apparent in practice. As in our previous examples, then, conceptual practice has to be seen as *temporally emergent,* as do its products.[26] Likewise, it is appropriate to note the *posthumanist* aspect of conceptual practice as exemplified in Hamilton's work. My analysis again entails a decentering of the human subject, though this time toward disciplinary agency rather than the material agency that has been at issue in earlier discussions. Here, once more, it is not the case that Hamilton as a human agent disappears from my analysis. I have not sought to reduce him to an "effect" of disciplinary agency, and I do not think that that can sensibly be done. It is rather that the center of gravity of my account is positioned *between* Hamilton as a classical human agent and the disciplines that carried him along. To be more precise, at the center of my account is the dance of intertwined human and disciplinary agency that traced out the trajectory of Hamilton's practice.[27]

So far I have been talking about the transformation of modelling vectors and formalisms in conceptual practice. But more was mangled and interactively stabilized in our example than that, and I want to talk first about the intentional structure of Hamilton's work before returning to its disciplinary aspects. One must, I think, take seriously Hamilton's already quoted intention "to connect . . . *calculation* with *geometry,*

26. Discussing Hamilton's unsuccessful attempts at constructing triplet systems, historians often invoke later mathematical existence proofs that appear to be relevant. For example, Hankins (1980, 438 n. 2) reproduces the following quotation from the introduction to volume 3 of Hamilton's collected papers (Hamilton 1967, xvi): "Thirteen years after Hamilton's death G. Frobenius proved that there exist precisely three associative division algebras over the reals, namely, the real numbers themselves, the complex numbers and the real quaternions." One is tempted to conclude from such assertions that Hamilton's search for triplets was doomed in advance (or fated to arrive at quaternions) and that the temporal emergence of his practice and its products is therefore only apparent. Against this, one can note that proofs like Frobenius's are themselves the products of sequences of practices that remain to be examined. There is no reason to expect that analysis of these sequences would not point to the temporal emergence of the proofs themselves. Note also that these sequences were precipitated by Hamilton's practice and by subsequent work on triplets, quaternions, and other many-place systems, all of which served to mark out what an "associative division algebra over the reals" might mean. Since this concept was not available to Hamilton, he cannot have been looking for new instances of it. On the defeasibility of "proof," see Lakatos 1976; and Pinch 1977.

27. Since the discussions of temporal emergence and posthumanism in the earlier chapters, like those in the literature more generally, are concerned with situations of cultural heterogeneity and the interplay of human and nonhuman agency, it is worth noting explicitly that the emergent and posthumanist aspects of Hamilton's conceptual practice are *not* tied to any degree of cultural heterogeneity. Unlike my previous examples, Hamilton's practice moved in a homogeneous (though multiple) space of conceptual resources and disciplines.

through some *extension,* to *space of three dimensions.*" One cannot otherwise make sense of the dialectics of resistance and accommodation that steered his practice through the open-ended space of modelling and eventually terminated in the quaternion system. An important point to note, however, is that Hamilton's goal was radically mangled. He aimed at an association in three dimensions, but he actually achieved one in four, via the shift in bridgehead implicit in the introduction of the new square root of -1 that he called k.

Two comments are appropriate here. First, that this is a further instance of what we saw before in the history of Glaser and the bubble chamber (chapter 2), namely, that goals and purposes are *in the plane* of scientific practice. They are not entities that control it from without. Thus Hamilton's goal was conceivable only within the cultural space in which an association between complex algebra and geometry had already been constructed, and it was further transformed (to an orientation to four instead of three dimensions) in the real time of his practice, as part and parcel of the dialectic of resistance and accommodation that we have examined. There is, though, something of a difference between this mangling of Hamilton's goal and Glaser's. Unlike Glaser, it does not seem that Hamilton self-consciously changed his goal in his final steps toward the quaternion system. There is no reason to think that he first decided to try to construct a four-place formalism and then introduced the new quantity k as a bridgehead en route. It is better, I believe, to think of the introduction of k as an immediate technical accommodation to the resistance that had arisen in Hamilton's technical practice. Only after working through this move and arriving at the quaternion system was it apparent that Hamilton had achieved a goal quite different from that at which he had aimed. In this instance, then, the mangling of the intentional structure of human agency amounted, if only temporarily, to its *effacement,* its vanishing, in the exigencies of the moment.

We can turn now from the intentional structure of human agency to its disciplined, repetitive, machinelike aspects. I have emphasized that Hamilton was carried along in his practice by disciplinary agency, and it was crucial to my analysis that in his transcriptions he acted without discretion. Such lack of discretion is the precondition for the emergence of dialectics of resistance and accommodation. But it is worth emphasizing also that Hamilton evidently did exercise discretion in choosing just which disciplines to submit himself to. Thus, throughout his practice, he maintained the first part of the geometrical rule already established for the multiplication of lines in the complex plane (that the length of the product of two line segments is the product of their individual lengths). But it was crucial to his path to quaternions that at a certain point he simply abandoned the second part of the multiplication rule concerning the orientation of product lines in space. He did not attempt to transcribe this when thinking about the multiplication of two arbitrary triplets. Part of Hamilton's strategy of accommodation to resistance was, then, a selective and tentative modification of discipline—in this case, an *eliminative* one. Hamilton bound himself to a part but not all of established routine practice.[28]

One can also understand Hamilton's introduction of noncommuting quantities into his extension of complex algebra as a selective modification of discipline, but in this case an *additive* one. He continued to follow standard practice as far as ordinary numbers were concerned—treating their products as indifferent to the order of terms to be multiplied—but invented a quite new and nonroutine rule for the multiplication of his various square roots of -1. In such ways, Hamilton both drew upon established routines to carry himself along and, as part of his accommodation to resistance, transformed those routines, eliminating or adding to them as seemed to him promising. Disciplinary agency, I therefore want to say, has again to be seen as in the plane of practice and mangled there in the very dialectics of resistance and accommodation to which it gives structure. And, further, transformed disciplines are themselves interactively stabilized in the achievement of cultural associations. That certain specific transformations of discipline rather than others should have been adopted was itself determined in the association of calculation with geometry that Hamilton eventually achieved with quaternions.

The theme of discipline will reappear in chapter 5, but to conclude the present discussion, I should connect it back to that of section 3. 5. There I analyzed the construction of an autochthonous discipline proper to the performance of a specific scientific experiment, while here I have been talking about the transformation of detached disciplines in a passage of mathematical practice. Despite this dissimilarity, I want to emphasize what the two analyses have in common. They demonstrate, I take it, that disciplines should be seen as yet more elements of the multiple and heterogeneous culture of science. And even when they become

28. Similarly, in his earlier attempts to construct an algebraic system of triplets modelled on his system of couples, in the face of resistances Hamilton abandoned the established algebraic principle of unique division (1853, 129–31).

detached from specific local circumstances and take on the quality I have ascribed to disciplinary agency, still they remain in the plane of practice, subject to interactive stabilization, destabilization, and restabilization alongside all of the other elements of scientific culture—material, conceptual, social—in a decentered process of temporally emergent mangling.

4.5 SCIENCE AND THE MANGLE

Apart from a postscript to this chapter, my empirical examination of scientific practice is now more or less complete. The next chapter is about technology and the industrial workplace. So here I should review what has been accomplished. First, I can note that the present chapter redeems the promise made in chapter 3. There I suggested that a full understanding of the mangling that sutures together the material and the conceptual in scientific knowledge production had to depend upon an understanding of the dynamics of purely conceptual practice, and that is what I have just sketched out. Once one grasps the idea that disciplinary agency endows conceptual practice with a temporally emergent structure, one can immediately appreciate, from the conceptual as well as the machinic end, how conceptual structures and machinic performances can fail to cohere under cultural extension, and how the material-conceptual dialectics of resistance and accommodation can arise that were centrally at issue in chapter 3. This means that I could summarize my overall analysis of science by simply adding what we have learned in this chapter to my brief summary at the end of chapter 3, but it might help to approach the task from a somewhat different angle this time.

In chapter 1, I stated my intention to develop an understanding of science in the performative idiom, an idiom capable of recognizing that the world is continually doing things and that so are we. Thus I have paid close attention to the machines and instruments that are integral to scientific culture and practice, and the conclusion of chapters 2 and 3 was that we should see the machinic field of science as being very precisely adjusted in its material contours to capture and frame material agency. The exact configuration of a machine or an instrument is the upshot of a tuning process that delicately positions it within the flow of material agency, harnessing and directing the latter—domesticating it. The image that lurks in my mind seems to be that of a finely engineered valve that both regulates and directs the flow of water from a pipe (though perhaps it is some kind of a turbine—I mentioned windmills in section 1.2). The performative idiom directs us also to think about human agency, and the argument that I have sought to exemplify in the last three chapters is that this can be grasped along similar lines. One should think about the scale and social relations of scientific agency, and the disciplined practices of such agency, as likewise being finely tuned in relation to its performativity. And, as I have been stressing all along, the engineering of the material and the human do not proceed independently of one another: in scientific culture, particular configurations of material and human agency appear as interactively stabilized against one another.

Once one begins to think about knowledge as well as performance, the picture becomes more elaborate, but its form remains the same. In chapter 3, my suggestion was that we should think of factual and theoretical knowledge in terms of representational chains passing through various levels of abstraction and conceptual multiplicity and terminating, in the world, on captures and framings of material agency. In this chapter, I have added that conceptual structures (scientific theories and models, mathematical formalisms) are themselves to be understood as positioned in fields of disciplinary agency, much as machines are positioned in fields of material agency. Conceptual structures are like precisely engineered valves, too, domesticating disciplinary agency. Again, though, conceptual engineering should not be thought of as proceeding independently of the engineering of the human and material. As exemplified in this chapter, disciplines, for instance, are themselves subject to transformation in conceptual practice, and, as exemplified in chapter 3, conceptual and machinic elements of culture evolve together in empirical practice. Scientific culture, then, appears as itself a wild kind of machine built from radically heterogeneous parts, a supercyborg, harnessing material and disciplinary agency in material and human performances, some of which lead out into the world of representation, of facts and theories.[29]

I have to confess that I like this image of scientific culture. It helps me to fix in my mind the fact that the specific contents of scientific

29. I have seen draft essays by John Law that evoke a related image of scientific and technological culture as a kind of giant plumbing system, continually under extension and repair.

knowledge are always immediately tied to specific and very precisely formed fields of machines and disciplines. Above all, it helps me to focus on the fact that scientific knowledge is just one part of the picture, not analytically privileged in any way but something that evolves in an impure, posthuman dynamics together with all of the other cultural strata of science—material, human, social (in the next section, I throw metaphysical systems into this assemblage, too, and in section 6.4, I add epistemic rules of method). This is, of course, in contrast with the representational idiom, which can hardly get the nonrepresentational strata of science into focus and which can never grasp its performative aspect.

I turn now to the question of how the supercyborg of scientific culture is extended in time. Traditional answers assert that something substantive within scientific culture (as I define it) endures through cultural extension and explains or controls it—social interests, epistemic rules, or whatever. Or perhaps something quite outside culture has the controlling role: the world itself. I have already criticized the idea that the social can play the required explanatory role (section 2.5), and I continue this argument in sections 4.6 and 5.3. In chapter 6, I also argue against any necessarily controlling role for epistemic rules, and I give my own account of how "the world itself" plays into cultural extension. For the moment, I want to stress that on my analysis *nothing* substantive explains or controls the extension of scientific culture. Existing culture is the surface of emergence of its own extension, in a process of open-ended modelling having no destination given or knowable in advance. Everything within the multiple and heterogeneous culture of science is, in principle, at stake in practice. Trajectories of cultural transformation are determined in dialectics of resistance and accommodation played out in real-time encounters with temporally emergent agency, dialectics that occasionally arrive at temporary oases of rest in the achievement of captures and framings of agency and of associations between multiple cultural extensions. I have noted, it is true, that one needs to think about the intentional structure of human agency to understand this process; vectors of cultural extension are tentatively fixed in the formulation of scientific plans and goals, and resistances have to be seen as relative to such goals. But as I have shown, plans and goals are both emergent from existing culture and at stake in scientific practice, themselves liable to mangling in dialectics of resistance and accommodation. They do not endure through, explain, or control cultural extension.

So this is my claim: there is no substantive explanation to be given for the extension of scientific culture. There is, however, and this is also my claim, a temporal pattern to practice that we can grasp, that we can find instantiated everywhere, and that constitutes an understanding of what is going on. It is the pattern just described—of open-ended extension through modelling, dialectics of resistance and accommodation, and so on. And in good conscience, this pattern—the mangle—is the only explanation that I can defend of what scientific culture becomes at any moment: of the configuration of its machines, of its facts and theories, of its disciplines and social relations, and so forth. Science mangles on.

4.6 POSTSCRIPT: MATHEMATICS, METAPHYSICS, AND THE SOCIAL

The central task of this chapter has been to understand how dialectics of resistance and accommodation can arise in conceptual practice. I want to end, though, by developing two subsidiary topics. It is common knowledge, among historians of mathematics at least, that Hamilton was as much a metaphysician as a mathematician, and that he felt that his metaphysics was, indeed, at the heart of his mathematics (Hankins 1980; Hendry 1984). I therefore want to see how the relation between mathematics and metaphysics can be understood in this instance. At the same time, it happens that David Bloor has offered a clear and interesting SSK-style explanation of Hamilton's metaphysics. So, as promised in my discussion of social theory in section 2.5, I can take this as an opportunity to characterize SSK in its application to historical examples and to clarify just where the mangle diverges from it. I start with Bloor and SSK.

Bloor's essay "Hamilton and Peacock on the Essence of Algebra" (1981; page citations below are to this work) focuses on the different metaphysical understandings of algebra articulated by Hamilton on the one side and a group of Cambridge mathematicians including Peacock on the other. We can get at this difference by returning to the foundational crisis in nineteenth-century algebra. In section 4.2, I suggested that the geometrical representation of complex algebra was one way of defusing the crisis and giving meaning to negative and imaginary quantities. But various mathematicians did not opt for this commonsense route, preferring more metaphysical approaches. Peacock and the Cambridge mathematicians took a *formalist* line, as Bloor calls it, which suggested that mathematical symbols and the systems in which they were embedded were sufficient unto themselves, in need of no extra-mathematical foundations and subject to whatever interpretation proved appropriate to specific uses. Thus, from the formalist point of view, there was and could be no foundational crisis in algebra. A quotation from the mathematician George Boole sums up this position nicely.

> They who are acquainted with the present state of the theory of Symbolical Algebra, are aware, that the validity of the processes of analysis does not depend upon the interpretation of the symbols which are employed, but solely upon the laws of their combination. Every system of interpretation which does not affect the truth of the relations supposed, is equally admissible, and it is thus that the same process may, under one scheme of interpretation, represent the solution of a question on the properties of numbers, under another, that of a geometrical problem, and under a third, that of a problem of dynamics or optics. (Boole 1847, 1, quoted by Nagel 1979, 166)

Hamilton disagreed. He thought that mathematical symbols and operations must have some solid foundations that the mind latches onto—consciously or not—in doing algebra. And, as Bloor puts it: "Hamilton's metaphysical interests placed him securely in the Idealist tradition. He adopted the Kantian view that mathematics is synthetic *a priori* knowledge. Mathematics derives from those features of the mind which are innate and which determine *a priori* the general form that our experience must take. Thus geometry unfolds for us the pure form of our intuition of space. Hamilton then said that if geometry was the science of pure space, then algebra was the science of pure time" (204). And, indeed, Hamilton developed his entire theory of complex algebra explicitly in such terms. In his "Theory of Conjugate Functions, or Algebraic Couples: With a Preliminary and Elementary Essay on Algebra as the Science of Pure Time," first read to the Royal Irish Academy in 1833 (Hamilton 1837), he showed how positive real algebraic variables—denoted *a, b, c,* and so forth—could be regarded as "steps" in time (rather than magnitudes of material entities), and how negative signs in front of them could be taken as denoting reversals of temporality, changing before into after. He also elaborated the system of couples mentioned in section 4.2. Written (a, b), these couples transformed like the usual complex variables under the standard mathematical operations but, importantly, the problematic symbol i was absent from them. Hamilton's claim was thus to have positively located and described the foundations of complex algebra in our intuitions of time and its passing.[30]

This much is well known, but Bloor takes the argument one step further: "I am interested in why men who were leaders in their field, and who agreed about so much at the level of technical detail, nevertheless failed to agree for many years about the fundamental nature of their science. I shall propose and defend a sociological theory about Hamilton's metaphysics and the divergence of opinion about symbolical algebra to which he was a party" (203). Bloor's idea is thus, first, that the technical substance of algebra did not determine its metaphysical interpretation, and therefore, second, that we need to invoke something other than technical substance—namely, the social—to explain why particular individuals and groups subscribed to the metaphysical positions that they did. This is a standard opening gambit in SSK, and Bloor follows it up by discussing the different social positions and visions of Hamilton and the Cambridge formalists and explaining how particular metaphysical views serve to buttress them.[31] According to Bloor, Hamilton was aligned with Coleridge and his circle and, more broadly, with "the interests served by Idealism" (220)—conservative, holistic, reactionary interests opposing the growing materialism, commercialism, and individualism of the early nineteenth century and the consequent breakdown of the traditional social order. As Bloor explains it, Hamilton's idealism assimilated mathematics to the Kantian category of understanding. Understanding in turn was understood to be subordinate to the higher faculty of reason. And, on the plane of human affairs, reason was itself the province not of mathematics but of religion and the church. Thus the "practical import" of Hamilton's idealism "was to place mathematics as a profession in a relation of general subordination to the Church. Al-

30. Hendry offers a subtle analysis of the early development of Hamilton's Kantianism and suggests that Hamilton might possibly have developed his theory of couples independently of it, hitching the metaphysics to the algebra "as a vehicle through which to get the essay published" (1984, 64). I should note that this observation creates no problems for my argument in what follows. I do not insist that practice always has a metaphysical flavor, but I do want to insist that metaphysics, when relevant, is subject to change and transformation in practice.

31. The only respect in which Bloor's essay is untypical of SSK is that he stops short, explaining metaphysics without pressing on into the technical substance of science. He does remark, however, that "should it transpire that this metaphysics is indeed relevant to technical mathematics, then my ideas may help to illuminate these matters as well" (206). I am more concerned with the overall form of Bloor's argument than with its restriction to metaphysics.

gebra, as Hamilton viewed it, would always be a reminder of, and a support for, a particular conception of the social order. It was symbolic of an 'organic' social order of the kind which found its expression in Coleridge's work on Church and State" (217).

So Hamilton's social vision and aspirations structured his metaphysics. As far as the Cambridge group of mathematicians was concerned, the same pattern was repeated, but starting from a different point. In mathematics and beyond, they were both "reformers and radicals" (222) and "professionals" (228) keen to assert their autonomy from traditional sources of authority like the church. Their formalism and its opposition to the foundationalism of people like Hamilton, then, served this end, defining mathematics as the special province of mathematicians. It was an antimetaphysics, one might say, which served to keep metaphysicians and the church out. Thus Bloor's analysis of the differences between the two parties over the foundations of mathematics. As he summarizes it: "Stated in its broadest terms, to be a formalist was to say: 'we can take charge of ourselves.' To reject formalism was to reject this message. These doctrines were, therefore, ways of rejecting or endorsing the established institutions of social control and spiritual guidance, and the established hierarchy of learned professions and intellectual callings. Attitudes toward symbols were themselves symbolic, and the messages they carried were about the autonomy and dependence of the groups which adopted them" (228).

I have no quarrel with Bloor's arguments as rehearsed so far. I have no knowledge of the social locations and aspirations of the parties concerned that would give me cause to doubt the existence of the social-metaphysical-mathematical correlations he outlines. But still, something peculiar happens toward the end of Bloor's essay. He concludes by stating that "I do not pretend that this account is without problems or complicating factors" (228) and then lists them. For the rest of this section, we will be concerned with just one complicating factor.

"It is necessary," Bloor remarks, "to notice and account for the fact that Hamilton's opposition to Cambridge formalism seemed to decline with time. In a letter to Peacock dated Oct. 13, 1846, Hamilton declared that his view about the importance of symbolical science 'may have approximated gradually to yours.' Interestingly, Hamilton also noted some four years later 'how much the course of time has worn away my political eagerness'" (229). The structure of these sentences is, I think, characteristic of what SSK looks like when brought to bear upon empirical studies. Note first that the shift in Hamilton's metaphysics is viewed as a "problem." It appears this way to Bloor because he wants to understand the social as not just a correlate of the metaphysical but as a kind of cause.[32] The social is the solid, reliable foundation that holds specific metaphysical positions in place in an otherwise open-ended space. Any drifting of Hamilton's metaphysics threatens this understanding, and Bloor therefore tries to recoup this drift, by qualifying it as perhaps apparent—"seemed to decline"—and then by associating it with a decline in Hamilton's "political eagerness." Perhaps Hamilton's social situation and views changed first and gave rise to Hamilton's concessions to formalism, seems to be Bloor's message (though the dates hardly look promising). If so, Bloor's causal arrow running from the social to the metaphysical would be secure.[33]

In what follows, I want to offer a different interpretation of Hamilton's metaphysical wandering, but before that I want to comment further on Bloor's general position. Three points bear emphasis. First, although I have hitherto described SSK as tending to regard the social as a nonemergent cause of cultural change in science, it is clear that Bloor *does* recognize here that the social can change with time. This is precisely how he hopes to cope with the problem of changing metaphysics. But second, he offers no examination or analysis of how the social changes. The social, I want to say, is treated as an at most *quasi-emergent* category, both in this essay and in the SSK canon in general.[34] The gaze of SSK only catches a fixed image of the social in the act of structuring the development of the technical and metaphysical strata of science. SSK always seems to miss the movie in which the social is itself transformed.[35] Bloor's essay, then, exemplifies the difference between SSK and the mangle that I mentioned in section 2.5: in contrast to SSK, I have

32. Bloor (1991, 7) lists causal social explanation as the first distinguishing mark of the strong program in SSK.

33. Thus Bloor's text immediately following the previous quotation continues: "A corresponding and opposite movement took place in Whewell's life. Here, *in obliging conformity with my thesis,* it is known that as Whewell moved to the right . . . he increasingly moved away from the symbolical approach in his mathematical writings" (229–30, emphasis added).

34. One can treat this quasi-emergent vision of the social as definitional of the SSK canon. It is not a circular definition since it largely coincides with at least the early key texts that most authors in science studies associate with SSK (see chap. 1, n. 4).

35. In its early development, SSK was articulated against philosophical positions that rancorously opposed the suggestion that there was *anything* significantly social about scientific knowledge. A concentration on situations where the social could plausibly be regarded as both fixed and as explanatory of metaphysical and technical developments therefore fulfilled a strategic argumentative function for SSK. SSK's endless deferral of any

already argued that the social should in general be seen as in the plane of practice, both feeding into technical practice and being emergently mangled there, rather than as a fixed origin of unidirectional causal arrows. Third, it is characteristic of SSK that Bloor does not even consider the possibility that there might be any explanation for Hamilton's metaphysical shift *other than* a change in the social. In contrast, I now want to offer an explanation that refers this shift not outward to the social but inward, toward Hamilton's technical practice.

Bloor says that Hamilton's opposition to formalism "seemed" to decline, but the evidence is that there was no "seeming" about it. As Hamilton put it in another letter written in 1846 (to his friend Robert Graves): "I feel an increased sympathy with, and fancy that I better understand, the Philological School [Bloor's formalists]. It enables me to see better the high functions of language, to trace more distinctly and more generally the influence of signs over thoughts, and to understand an answer which I hazarded some years ago to a question of yours, What did I suppose to be the *Science of Pure Kind*? namely, that I supposed it must be the *Science of Symbols*" (quoted in Nagel 1979, 189).

In fact, 1846 seems to be an important date in Hamilton's metaphysical biography. It was just around then that he began to indicate in various ways that his position had changed. One might suspect, therefore,

---

inquiry into how the social might itself evolve seems strange, though, even given that background. It is, I suspect, part and parcel of SSK's almost principled refusal to interrogate key sociological concepts like "interest." Thus, in 1977, we find Barry Barnes writing in *Interests and the Growth of Knowledge* that "new forms of activity arise not because men are determined by new ideas, but because they actively deploy their knowledge in a new context, as a resource to further their interests" (78), but then, on the last page, he shuffles interests out into unexplored regions of social theory with the remark that "I have deliberately refrained from advancing any precise definitions of 'interest' and 'social structure'; this would have had the effect of linking the claims being advanced to particular schools of thought within sociological theory. Instead, I have been content, as it were, to latch the sociology of knowledge into the ongoing general trends of social thought" (86). Nothing has changed in the intervening years. Interests, and de facto their dynamics, are still left out in the cold by SSK. In a recent essay review of Latour's *Science in Action,* Shapin writes: "One must . . . welcome any pressure that urges analysts further to refine, define, justify and reflect upon their explanatory resources. If there is misunderstanding, by no means all the blame needs to be laid at Latour's door. 'Interest-explanation' does indeed merit further justification" (1988a, 549). And replying to his critics in the afterword to the second edition of *Knowledge and Social Imagery,* Bloor writes that "[u]ndeniably the terminology of interest explanations is intuitive, and much about them awaits clarification" (1991, 171). In chapter 3 of that same book, Bloor advances the Durkheimian argument that resistance to the strong program arises from a sacred quality attributed to science in modern society. Perhaps in SSK it is the social that has become sacred.

that Hamilton's worries about metaphysical idealism had their origins in his technical practice around quaternions in the early 1840s.[36] And this suspicion is supported by the fact that Hamilton's technical writings on quaternions—specifically the preface to his first book on the subject, the massive *Lectures on Quaternions* (1853)—contain several explicit discussions of his past and present metaphysical stances. We can peruse a few and try to make sense of what happened.

Hamilton's preface to the *Lectures* takes the form of a historical introduction to his thought and to related work of other mathematicians, and one striking feature of it is the tone of regret and retraction that Hamilton adopts whenever the Science of Pure Time comes up. The preface begins with a summary of his early work on couples, which he introduces with the remark that

> [i]n this manner I was led, many years ago, to regard Algebra as the SCIENCE OF PURE TIME . . . If I now reproduce a few of the opinions put forward in that early Essay, it will be simply because they may assist the reader to place himself in that *point of view,* as regards the first elements of *algebra,* from which a passage was gradually made by me to that comparatively *geometrical* conception which it is the aim of this volume to unfold. And with respect to anything unusual in the *interpretations* thus proposed, for some simple and elementary notations, it is my wish to be understood as not at all insisting on them as *necessary,* but merely proposing them as consistent amongst themselves, and preparatory to the study of quaternions, in at least one aspect of the latter. (117–18)

So much for a priori knowledge.

Later, Hamilton verges upon apology for mentioning his old metaphysics.

> Perhaps I ought to apologize for having thus ventured here to reproduce (although only historically . . .) a view so little supported by scientific authority. I am very willing to believe that (though not unused to calculation) I may have habitually attended too little to the *symbolical* character

36. Hankins (1980, 310) briefly connects Hamilton's metaphysical shift with his technical practice along the lines elaborated below. I should mention that Hamilton's Kantianism had a second string besides his thinking about time—namely, a concern with triadic structures grasped in relation to the Trinity (Hankins 1980, 285–91). Besides possible utility, then, Hamilton's searches for triplets and his concern with three-dimensional geometry have themselves a metaphysical aspect. My focus here, though, is with his overall move away from Kantianism toward formalism.

> of Algebra, as a Language, or organized system of *signs:* and too much (in proportion) to what I have been accustomed to consider its *scientific* character, as a Doctrine analogous to Geometry, through the Kantian parallelism between the *intuitions* of Time and Space. (125)

Later still, Hamilton speaks positively about the virtues of formalism and their integration into his own mathematical practice, saying that he "had attempted, in the composition of that particular series [of papers on quaternions understood as quotients of lines in three-dimensional space, published between 1846 and 1849], to allow a more prominent influence to the general *laws of symbolical language* than in some former papers of mine; and that to this extent I had on this occasion sought to imitate the *Symbolical Algebra* of Dr. Peacock" (153).

Far from being situated on the opposite side of a metaphysical gulf from Peacock, then, by 1846 Hamilton was *imitating* Peacock's formalist approach in his technical practice (without, I should add, entirely abandoning his earlier Kantianism). And to understand why, we need, I think, to look more closely at that practice. In the very long footnote that begins with the apology for mentioning the Science of Pure Time, Hamilton actually goes on to assert that he could have developed many of the aspects of the quaternion system to be covered in the rest of the book within his original metaphysical framework, and that this line of development "would offer no result which was not perfectly and easily *intelligible,* in strict consistency with that *original* thought (or intuition) of time, from which the whole theory should (on this supposition) be evolved." (125). "Still," he continues,

> I admit fully that the actual *calculations* suggested by this [the Science of Pure Time], or any other view, must be performed according to some fixed *laws of combination of symbols,* such as Professor De Morgan has sought to reduce, for ordinary algebra, to the smallest possible compass . . . and that in following out such *laws* in their symbolical consequences, uninterpretable (or at least uninterpreted) *results* may be expected to arise. . . [For example] in the passage which I have made (in the Seventh Lecture), from *quaternions* considered as *real* (or as geometrically *interpreted*), to *biquaternions* considered as *imaginary* (or as geometrically *uninterpreted*), but as symbolically *suggested* by the generalization of the quaternion formulae, it will be perceived . . . that I have followed a *method of transition,* from *theorems proved* for the *particular* to *expressions assumed* for the *general,* which bears a very close *analogy* to the methods of Ohm

and Peacock: although I have *since* thought of a way of *geometrically interpreting the biquaternions* also. (125–26)

Now, I am not going to exceed my competence by trying to explain what biquaternions are and how they specifically fit into the story, but I think one can get an inkling from this quotation of how and why Hamilton's metaphysics changed. While Hamilton had found it possible calmly to work out his version of complex algebra on the basis of his Kantian notions about time, in his subsequent mathematical practice leading through quaternions he was, to put it crudely, flying by the seat of his pants. He was struggling through dialectics of resistance and accommodation, reacting as best he could to the exigencies of technical practice, without much regard to or help from any a priori intuitions of the inner meanings of the symbols he was manipulating. The variety of the bridging and filling moves that he made on the way to quaternions that I reviewed in section 4.3, for example, hardly betray any "strict consistency" with an "*original* thought (or intuition)." Further, what guided Hamilton through the open-ended space of modelling was, I argued, disciplinary agency—the replaying of established *formal* manipulations in new contexts marked out by bridging and filling. And, at the level of products rather than processes, a similar situation obtained. Hamilton continually arrived at technical results and then had to scratch around for interpretations of them—starting with the search for a three-dimensional geometric interpretation of his initial four-dimensional formulation of quaternions, and ending up in the quotation just given with biquaternions ("I have *since* thought of a way"). Moreover, Hamilton proved to be able to think of several ways of interpreting his findings. In the preface to the *Lectures,* he discusses three different three-dimensional geometrical interpretations, one of which (not that mentioned at the end of section 4.3) forms the basis for his exposition of quaternions in the book itself (145–54).[37] Formal results followed by an indefinite number of interpretations: this is a description of formalist metaphysics.

So there is a prima facie case for understanding the transformation

37. Of one of these systems, Hamilton wrote, "It seemed (and still seems) to me natural to connect this *extra-spatial unit* [the nongeometrical part of the quaternion] with the conception of TIME." But then he reverted to the formalist mode: "Whatever may be thought of these abstract and semi-metaphysical *views,* the *formulae* . . . are in any event a sufficient *basis* for the erection of a CALCULUS of quaternions" (152).

in Hamilton's metaphysics in the mid–1840s as an accommodation to resistances arising in technical-metaphysical practice. A tension emerged between Hamilton's Kantian a priorism and his technical practice, to which he responded by attenuating the former and adding to it an important dash of formalism. My suggestion is, therefore, that we should see metaphysics as yet another heterogeneous element of the culture that scientists operate in and on. Like the technical culture of science, like the social, and like discipline, metaphysics is itself at stake in practice, and just as liable to temporally emergent mangling there in interaction with all of those other elements. That is the positive conclusion of this section as far as my analysis of practice is concerned.

Comparatively, I have tried to show how my analysis differs from SSK in its handling of a specific example. Where SSK necessarily looks outward from metaphysics (and technical culture in general) to quasi-emergent aspects of the social for explanations of change (and stability), I have looked inward, to technical practice itself. There is an emergent dynamics there that goes unrecognized in SSK. I have, of course, said nothing on my own account about the transformation in Hamilton's "political eagerness" that Bloor mentions. Having earlier argued for the mangling of the social, I find it quite conceivable that Hamilton's political views might also have been emergently mangled and interactively stabilized alongside his metaphysics in the evolution of the quaternion system. On the other hand, they might not. I have no more information on this topic than Bloor—but at least the mangle can indicate a way past the peculiar quasi-emergent vision of the social that SSK offers us.

FIVE

Technology

Numerically Controlled Machine Tools

In this last chapter of part 1, I move from science to technology and the industrial workplace. In our society, industrial machinery is the site of capture of material agency par excellence, and by focusing on this topic I want to establish empirically what should already be clear in the abstract: while the representational idiom necessarily registers a sharp discontinuity in the move to technology—industrial machines are directly oriented to material performance, not to knowledge and representation—the performative idiom, in contrast, carries smoothly over into the analysis of technology and production.[1] I hope to show, then, how my analysis of the mangle speaks directly to the concerns not just of science studies but of the broader field of science, technology, and society (STS) studies.

Beyond extending the range of my analysis, the present chapter also extends and elaborates upon themes and arguments set out in earlier chapters. Thus the case study that follows sheds light on *the mangling of the social,* here meaning social performances and relations, and thus continues arguments begun in section 2.4. Further, the performances and social relations centrally at issue are those of shop-floor labor, as

1. As Simon Schaffer pointed out to me, this chapter could well have been the first rather than the last of part 1. From the perspective of the performative idiom, the scientific machines and instruments discussed in chapters 2 and 3 appear a little odd in their close relation to the production of knowledge and representations. It would therefore make sense to begin my exposition with machines that capture material agency for its own sake, as it were. Biographically, however, I began my analyses of practice with the intention of getting to grips with science, and the center of gravity of my studies remains there, so I prefer to leave this opening into technology until the end.

imposed by management in explicit disciplinary codes. I will therefore be able in this chapter to take further the discussion of discipline and its mangling, begun in section 3.5 and carried on in chapter 4. Finally, though the industrial workplace has been strangely ignored by recent initiatives in technology studies, it is, of course, one of the classic sites for the development of social theory, and at the end of this chapter I will take the chance to exemplify the divergence between my analysis of cultural extension and more traditional sociological accounts, thus carrying on from where I left off in sections 2.5 and 4.6.[2]

5.1 NUMERICALLY CONTROLLED MACHINE TOOLS

My example is taken from *Forces of Production* (1986), David Noble's wonderful study of the development of numerically controlled machine tools and their introduction into the workplace. Specifically, I am interested in the latter, as discussed in chapter 11 of Noble's book, but some preliminary remarks are called for first. To begin at the beginning, a machine tool is a device to cut metal to precise specifications—a lathe or a milling machine or something of that ilk. It is clearly, to my way of thinking, a prototypical device for capturing nonhuman agency: one can accomplish things with a lathe that naked human agency could never accomplish. However, it is important to note that traditional machine tools do not cut metal of their own volition. They need a skilled operator to channel their agency in desired directions. This does not undercut my idea that material machines capture nonhuman agency; it implies, rather, that what matters in metal cutting is a human-machine couple—the lathe and its skilled operator come together as a single unit of machinic capture; in industrial production, they constitute a composite

2. The recent initiatives I have in mind are the "social construction of technological systems" (SCOTS) approach laid out by Pinch and Bijker, and the actor-network approach (on SCOTS, see Pinch and Bijker 1984 and Bijker, Hughes, and Pinch 1987; on the actor-network, see the works already cited by Callon, Latour, and Law; for an important recent collection of essays on technology, see Bijker and Law 1992). SCOTS studies have tended to focus on the social aspects of technology design rather than implementation. Actor-network studies have been interested in implementation, but largely outside the factory. On a theoretical plane, my approach departs from SCOTS inasmuch as in its early articulations the latter sought to transplant non- or quasi-emergent SSK-style frameworks from science to technology, though conversations with Trevor Pinch have persuaded me that such is no longer necessarily the case. For a point of divergence with the actor-network, see note 11 below. The industrial workplace remains, of course, a key arena for Marxist studies of technology and the social, and section 5.3 confronts the analysis offered by one of the leading authors of such studies.

human/nonhuman agent, a *cyborg*, to borrow Donna Haraway's (1991b) term (to which I return in section 7.2). And we can take the cyborg idea further. According to Noble, within the frame of twentieth-century corporate capital, little cyborgs composed of the conjunction of a single human being and a single machine have not been very important. Instead we find sociocyborgs: arrays of lathes and milling machines in a corporate machine shop, operated by wage labor within a classic Taylorite disciplinary apparatus of specified social roles and relations—a hierarchical command structure, precise job descriptions, production targets, rewards and penalties, and so forth.

So we have the idea of the Taylorite sociocyborg as the basic unit for the machinic capture of agency in the metalworking industries. And, as Noble tells the story, after World War II this sociocyborg was not seen by management as performing this capture very well.[3] Resistance, to use my term, relative to the goals of capital, was apparent. And one tentative accommodation to this resistance was developed for industry by engineers at MIT, in the shape of numerically controlled machine tools—N/C for short. The idea of N/C was to shift the balance of agency within the basic cyborg unit. Instead of requiring detailed human control, N/C equipment was controlled by digital computers executing instructions compiled by programmers; and, as a corollary, shop-floor labor was reduced to the role of button pushers. Or so, at least, it worked out in the Servomechanisms Lab at MIT. Now I want to concentrate on the implementation of N/C technology in industry. In particular, following Noble, I want to discuss the introduction of N/C at the General Electric Aero Engine Group plant at Lynn, Massachusetts, in the early 1960s (Noble 1986, chap. 11; page citations below are to this book), and I want especially to examine the manglings of the social that were consequent upon it. It is impossible to do justice here to the wealth of detail that Noble presents, but in outline the story goes like this.

The industrial implementation of N/C involved a double displacement: of the material technology from the laboratory to the shop floor, ripping out one set of machines and replacing them with another, and, at the same time, a reduction in pay rates of the human component of production (to R–17 from R–19 for conventional lathes in the company's classification—this last change being legitimated within the Taylorite frame of reference by the decrease in skill supposed to accompany

3. Here and subsequently I follow Noble in developing the narrative largely from the perspective of management and engineers rather than from that of shop-floor labor.

the new machines). One would expect from my earlier discussions of captures of agency that such displacements would not be unproblematic, that, in fact, the existing grip on productive agency would be loosened, not tightened, that resistances would arise, and that the reconfigured sociocyborg would need *tuning*. And so it proved. On arrival, N/C proved not to be the solution to GE management's problems. Quite the reverse.

In their thinking about N/C, GE management had implicitly adopted a technological-determinist scenario: the innate properties of the new material technology would themselves guarantee increased production of high-quality engine parts. Given that the N/C lathes simply executed instructions preprogrammed elsewhere—the shop-floor operators were not involved in this process—how else could it be? This scenario, however, failed to materialize in practice. It emerged, instead, that when the workers performed like the button pushers that management intended them to become, production rates dropped, as did the quality of finished parts produced (270–71; see also 221 and 244–46). At the same time,

> [t]he workers and the union protested strongly against the lower [pay] rate, insisting that, if anything, the N/C machines required more skill, not less, and that the rate should reflect this reality. The local argued that N/C demanded greater attention while the machine was in operation, in order to anticipate and correct for, or avoid, foul-ups, and that this required skill and experience and resulted in more tension and fatigue. Moreover, the union contended that, since tolerances were interrelated on some parts . . . meeting specifications had become more difficult. Finally . . . the machines could not simply run themselves, even with tape control, because frequent manual interventions during the tape cycle were necessary to check tolerances, make tool adjustments, compensate for tool wear and work piece irregularities, and otherwise insure a good finish. This was especially true when proving out new programs for new parts, when the chance of error hovered above 90 percent. (270)

Resistance to the introduction of N/C emerged, then, on the connected fronts of production rates and qualities and worker cooperation. And it is worth noting at once that this initial resistance to the introduction of N/C was truly temporally emergent in practice. It simply turned out in practice that the rate and quality of parts production dropped: under management's original description of N/C, there should have been no leeway for this. Likewise, a space for labor protest was created at the same time: if the introduction of N/C had succeeded in maintaining or improving production while reducing the workers to no more than button pushers, the rate cut would have been unchallengeable within the existing system of labor relations. In any event, in accommodation to this resistance, GE management employed several tactics at its disposal to try to improve N/C production, including under-the-table payments of higher rates as incentives to selected workers, but these succeeded only in further alienating the shop floor, and culminated in a major strike in January 1965, which was brought to a close only by an increase in the rate for N/C operators back to R–19 (275–76).[4]

Beyond ending the strike, this first accommodation of management to resistance failed. "The production problems remained, the unreliability, the programming errors, the excessive downtime, compounded by scheduling problems, worker and management turnover, and low morale" (276). Management then attempted a second accommodation within their traditional understanding of labor relations: "[They] kept increasing the pressure. Lead hands . . . were used . . . to increase the level of pacing. Supervisors bribed operators to enlarge output by promising them a lead-hand rate . . . Foremen also pitted the operators against each other in the competition for the higher rate" (277). Here we see Taylorite discipline being used as a tuning device, and the open-endedness of this process bears emphasis. As is evident from the above quotation, such mechanisms, and the disciplined roles and relations that they specify, should be seen as emergently adaptable to particular situations, with adaptation in this instance amounting to intensification.[5] But still, far from effecting the intended tuning of the sociocyborg of production, this intensification "led to a deep distrust and general breakdown in relations between the men on the floor and, most important, between operators and supervisors. Alienation and hostility had become the norm in Building 74. The part of the plant with the most sophisticated equipment had become the part of the plant with the highest scrap

4. That the GE workers took their struggle so far should also be seen as a temporally emergent feature of this episode. As one union researcher wrote, surveying the introduction of N/C equipment in other industries: "An analysis of the data collected thus far . . . reveals no established pattern. What happens to the wage rate in the transition from manual to automated appears to be tied in directly with the power and skill of negotiations of the locals involved" (274–75).

5. To align this comment with the analysis given in earlier chapters, I should note that existing implementations of Taylorism at GE constituted the *model* that was open-endedly extended in this instance.

rate, highest turnover, and lowest productivity, the 'bottleneck' in aircraft engine production" (277).

At this point, management shifted their tactics discontinuously. Instead of persisting with the attempt to accommodate resistance along traditional disciplinary lines, they relaxed discipline in favor of "job enrichment."[6] In the so-called Pilot Program, which came into operation in late 1968, many of the traditional disciplinary restraints on labor were withdrawn, and the N/C operators were encouraged to take over some, at least, of the functions of management in orchestrating production. The GE task force that laid the foundations for the Pilot Program argued that

> [t]he Task Force believes the principal reason for this lack of motivation [among N/C operators] is that management has been too steeped in traditional concepts of industrial engineering. These concepts . . . which served us well with older equipment and with the workforce of earlier generations seem to be at the source of our problems. It is the belief of the Task Force that the time has now come to break down many of the barriers that have existed for generations between job classifications and bar-

6. I thank Ted O'Leary for emphasizing to me that "job enrichment" initiatives were not particular to GE in this period (as indeed Noble's book makes clear). A recent essay by Peter Miller and Nikolas Rose (1993) provides important contextual information. They discuss a movement beginning in the 1950s that aimed "to improve the 'quality of working life' in the interests of the mental health and personal fulfillment of the worker, the ability and morality of the manager, the quality of the product, the efficiency and competitiveness of the enterprise and the political legitimacy of the corporation . . . Through participative design, worker representation, flexible hours, job enrichment, job enlargement, self-managed work teams, continual retraining and much else, work was to become democratic, creative, innovative and productive: a labor worthy of the modern self" (19). According to Miller and Rose, by the early 1970s, this had become a "self conscious international movement . . . receiving enthusiastic support not only from researchers, consultants, employers, and politicians, but also from such bodies as the International Labour Organization" (24). Although Miller and Rose do not discuss Noble's account of the Pilot Program, I think it is clear from what follows that the Pilot Program was conceived in relation to "quality of working life" concerns, and helped to specify what the intended new alignment of "the government of the workplace, the political problems of democracy and the ethics of subjectivity" (18) would amount to in practice (just as, for example, Luis Alvarez's bubble-chamber project helped to specify what big science would amount to in experimental-particle physics). An important early center for the movement was the Tavistock Institute of Human Relations, where the workplace was theorized in terms of "socio-technical systems" (20). The Tavistock researchers aimed at "a 'joint optimization' of the social and the technical . . . through one particular socio-technical device—that of the 'autonomous group'" and through a characterization of the organization "as an 'open system'" (22).

> gaining units. More succinctly, jobs need to be structured for the equipment and generation of workers of today. (280)[7]

Thus the new Pilot Program regime would be "unique in that there was to be no foreman, no scheduled lunch periods, and flexible starting and personal times" (281). Further, the classification of the operators would be "unlimited" (281), meaning that they would be free to start to take over responsibilities usually held by others, including management.

> The leaders (senior N/C machinists) . . . would [for example] assign N/C machinists in debugging new equipment, tools, and methods; schedule equipment start-up; work with planning in developing, implementing, and controlling new methods and procedures; approve programming from the viewpoint of good machine shop practice; review and make suggestions about changes in workstations, tools, and fixtures; assume responsibility for quality in the unit and interface with quality control; [and] monitor the area for availability of all materials and check equipment to insure safe and proper functioning. (280–81)

In short, in a drastic shift of policy, GE management now expected the pilots effectively to act like traditional management consultants and, moreover, to implement their own recommendations, blurring their roles into the traditional roles of foremen, planners, programmers, quality controllers, and so on.

One further aspect of the Pilot Program should be singled out for special attention. It was, by design, explicitly open-ended, in just the sense in which I have been using this term. As Noble puts it, "the program was an initiative taken by GE-Lynn management to *learn* how to achieve full use of N/C equipment by granting employees greater freedom and responsibility and eliciting from them knowledge about how best to process parts using such equipment" (281, emphasis added). Management, as it were, laid down the traditional reins of control, in the hope that the pilots could stabilize a new system of production centered on the N/C equipment.

> GE did not clarify what "pilot responsibilities" were, arguing repeatedly, to the point of near exasperation, that this was precisely what the Pilot Program was set up to find out. "Should we make the operators button-

7. Such lumping together of people and things—"the equipment and . . . workers of today"—is what leads Callon (1987) to refer to engineers as "engineer-sociologists" and to recommend their posthumanist sociology over the humanist sociologies of academics.

> pushers or responsible people?" "Should we make the tape right on the floor?" "What ought to be the operators' role in scheduling, maintenance, diagnostics, troubleshooting, inspection?" The company insisted that it did not know the answers to these questions, however uncomfortable that truth was for the participants and the union. "We don't know. We have not set any limits. They will start off the same as they are doing right now. The people will set the pace. They will determine what they will do. We don't know what the heck they will do. We are going to allow the people to make the determination as to how far they will go." "We are now involved," they concluded solemnly, "in what could be the most progressive step blue-collar workers have ever taken." (288)

And indeed, within the program, the pilots "seized the initiative."

> With management encouragement they became more involved in fixture repairs, tape debugging (working with programmers), diagnostics (working with maintenance personnel), inspection (working with quality control engineers), correcting planning sheets (working with planners), and scheduling (working with production engineers). In the context of such cooperation they developed new methods for tool orientation, and tooling changes, and new cutting paths on all drawings. In addition, they trained each other, especially those new to N/C, and gained an appreciation for the complexities and difficulties inherent in the production process. They came up with practical solutions to some of the problems. They recommended that there be runners for expediting the movement of material and tooling around the shop, that special attention be accorded to housekeeping, that the pilots be given their own cabinets for controlling the use of standard gauges as well as their own mini-tool crib to provide ready access to routinely used jigs and fixtures. In addition they requested that there be special set-up men and proposed more in-process inspection for the parts coming into their area from the roughing area . . . Finally, the pilots made suggestions about scheduling, how best to load the area in order to achieve the highest machine utilization. (289)

I have quoted this long passage to convey the extent of the open-ended evolution of work styles, roles, and relations, in the early phase of the Pilot Program. In 1970, though, the scope of the program was further expanded, allowing the pilots yet more control over the N/C work.

> "OK, we will let the pilots do everything and eliminate the MSO [manager of shop operations]." The men would now be allowed to "run their own job" in that the scope of their activities would be enlarged to include

> administration of vouchers, charting and evaluation of timekeeping . . . , veto on incoming parts from the roughing area . . . , and full processing responsibility (paperwork) for all MRB's (sent on to the military Materiel Review Board for inspection and approval). The lead hand would now function as quality control engineer, production scheduler, and distributor of work assignments . . . The pilots as a group would now do all administrative functions and paperwork, except discipline . . . , and therefore be in a position to take into account the effects of indirect costs on their performance and actually determine the best way of doing things, from their own perspective. (294–95)

The pilots forged ahead, then, into the open space ceded to them by GE management, devising new working arrangements as they went and progressively taking over the functions of management. And, according to Noble, the Pilot Program worked. The performance of the N/C shop improved; this tuning was successful (though, as Noble remarks on page 304, the union failed to keep its own records to demonstrate this fact). At the same time, however, new resistances emerged, not now centered on the N/C equipment itself, but instead around the edges of the Pilot Program, at the boundaries with the megacyborg of GE in its entirety. "Training remained woefully inadequate for the preparation of pilots for their new responsibilities . . . Conflicts with the support people in quality control, planning, and production also intensified, owing to the expanded responsibilities of the hourly workers in the Pilot Program, responsibilities which now seriously encroached upon jealously guarded territory" (295). Relations between GE corporate management and the local Lynn management also became strained. And, in response to these emergent resistances, one final accommodation was tried. In early 1975, GE management formally stated its decision to end the Pilot Program. N/C production at Lynn was reorganized on classic lines, with managerial functions stripped from the erstwhile pilots and returned to "planning, foremen, and supervisors" (317). Importantly, though, something had been learned in the experiment. "GE finally got everything it wanted: an N/C day rate restricted to R–19. . . , information on how most productively to use the new equipment, [and] a more flexible job description for operators" (313).

5.2 THE MANGLE AND THE SOCIAL

This is by no means the end of the story of N/C and the computerization of the workplace, at GE or elsewhere. The N/C shop at Lynn worked no better when Taylorism was brought back than it had in the first place, and the response of GE management was to look toward more computerization, to try even harder to squeeze the human out of the cyborg of production. But it is as far as I need to take the story. I want now to relate it to my overall picture of the mangle; then I can turn to Noble's own, rather different, analysis of the same happenings.

The story of N/C at GE is a story of the extension of sociomaterial culture, marked by a discontinuous shift in the realm of machines—from traditional machine tools to numerically controlled ones—and by progressive shifts in the realm of the social—shifts in job descriptions and in social relations among shop-floor labor and between labor and management. I have analyzed the latter in terms of a series of resistances and accommodations: as instances, that is, of mangling. And in this section, I want, as usual, to spell out the temporally emergent and posthuman aspects of this mangling of the social.

On emergence, it is clear that nothing in the pre-N/C culture of GE controlled or explained the twists and turns in the roles and relations of N/C production that we have been examining. To understand the oscillations between the early intensified disciplinary arrangements and the progressive blurring of workers' roles into those of management, to understand the actual course the Pilot Program took and the social contours of its end point as codified within the renewed disciplinary regime, one has to look to the specific resistances and accommodations that emerged as the implementation of N/C developed in time. No one, at GE or anywhere else, could have seen in advance what those resistances and accommodations would be. This is true in general, I think, but it is particularly clear in the evolution of the Pilot Program, in which both management and labor continually asserted (and, in the case of labor, complained) that there was no predetermined end point.

Turning to posthumanism, my suggestion is that the mangling of the social at GE cannot be understood in purely social (human) terms. Such an understanding cannot itself be centered on the social domain. And the easiest way to see this is to continue to follow Noble and to remain with the perspective of GE management on the N/C story. We can start by noting that one cannot ignore the intentions of GE management in the preceding narrative. Management plans and goals were clearly constitutive of events. Management held the initiative in deciding to install N/C equipment in the first place, in specifying different disciplinary arrangements at different times, and so on. I will talk more about GE management's goals in section 5.3, where I will argue that they should themselves be understood as posthumanly mangled. But an important (and related) point can be made prior to that discussion, namely, that the transformations of the social at GE cannot be reduced to management goals. The story cannot sensibly be made to revolve around them. Especially, I want to suggest, the material form and performance of the N/C machines themselves have to be seen as constitutive of the trajectory of emergence of work discipline at GE.

Four observations point to this conclusion. First, it is clear that, from the vantage point of GE management, the material form and performativity, as distinct from ideas and representations, of N/C mattered, in the sense that they were constitutive of the resistances encountered throughout the episode we have been discussing. As I noted earlier, according to prior expectations, there should have been no scope for resistance to N/C; it was only when the actual materials were in place that such scope emerged. Second, we can think about the role of computer programmers in the N/C story. In an earlier quotation, Noble mentioned the open-ended negotiation of relations between the pilots and the programmers, and, evidently, these relations could not even have existed apart from the material form of N/C; there was no place for programmers in pre-N/C production. Third, it is clear, I think, from Noble's description that the discipline that the pilots evolved within the Pilot Program—revised job descriptions and relations with other workers and with management—was an autochthonous one, progressively tuned to the material form and performance of the N/C equipment.[8] And fourth, this tuning constituted a learning process, the results of which GE management eventually recoded into the renewed Taylorite regime that superseded the Pilot Program.

The moral of these remarks is, then, that the story of the mangling of the social at GE again exhibits a posthumanist displacement. It does make sense to talk about the human agency and intentions of GE man-

8. Unfortunately, Noble does not go into sufficient detail to exemplify clearly an internal dialectic of resistance and accommodation within the Pilot Program. He simply states, in a passage quoted above, that the pilots "gained an appreciation for the complexities and difficulties inherent in the production process. . . [and] came up with practical solutions to some of the problems," which Noble lists (289). In the absence of reasons to think otherwise, however, I believe one should understand the evolution of new work disciplines within the program as structurally similar to the construction of the autochthonous laboratory disciplines discussed in section 3.5.

agement, but the center of gravity of the story lies elsewhere, at the point of intersection of human and material agency. The trajectory of evolution of the social has here to be understood in terms of emergent resistances and accommodations at the interface of these heterogeneous realms. And, I can add, although I have arrived at this conclusion from the perspective of management and their goals, it also follows if one starts from the human agency of shop-floor labor or from the material agency of N/C itself. Concerning the last, for example, we can note that the material contours and performativity of N/C did not *imply* any particular disciplinary transformations at GE. The twists and turns in discipline already discussed and the continuing instability of shop-floor culture at the cutoff point of my narrative are sufficient to establish that. It is also relevant to note both that the material form of N/C was itself at stake in this episode, at least to a limited extent, and that this was just one episode in an extended passage of management and engineering practice aimed at transforming the material and social elements of productive culture at once.[9] One cannot, in other words, make this story of the mangling of the social revolve around either human or material agency. What we have here is an example of the posthumanist decentering of the extension of sociomaterial culture.

Where does this leave us? Evidently one should not lean too heavily on a single example in generalizing about the endless interconnections between science, technology, and society. I have said nothing about the technical development of N/C within MIT, for example, or about the linkages that connected MIT outward into industry and vice versa.[10] Nevertheless, I want to suggest that we should take this example seriously, especially in its implication that, in society at large, technological developments are played out in an emergent and posthumanist fashion. In sections 7.3 and 7.4, I argue that we should see the overall development of STS in precisely these terms. For the moment, though, and to

9. Thus, under the Pilot Program, shop-floor workers were allowed to encroach upon the production of the computer tapes that ran the equipment, and the machines were "opened up" to allow labor to debug tapes in the real time of production (279–80, 288–89). As Noble makes clear, the N/C equipment installed at GE represented a single moment in the overall development of computerized production, and, as mentioned above, management initiatives at GE after the collapse of the Pilot Program took the form of yet more computerization and automation (322–23).

10. Much of Noble 1986 is devoted to the engrossing story of how the MIT Servomechanisms Laboratory took over (with military support) the development of N/C from industry, and then returned N/C equipment to industry in a drastically reconfigured form.

emphasize what is at stake in this idea, I can note how my analysis of technology-society relations differs from more traditional ones.[11]

Traditionally, the relations between technology and society have been understood in two opposed, but both temporally nonemergent, ways. The antihumanist version of the story is usually called *technological determinism,* and asserts that specific social transformations follow from given technological innovations, that the social is continually refashioned around technological imperatives (MacKenzie and Wajcman 1985). On the basis of the N/C example, it is clear, I think, that technological innovations can indeed have an impact on the social, acting as a shock to social relations, disciplines, and so forth, that had previously been tuned around earlier technology. Technology is not neutral, in this sense. Nevertheless, the argument of the previous paragraph but one undermines technological determinism: to note that no particular transformation of the social followed from N/C, and that the material form of N/C was itself at stake at GE, is to deny the autonomy and causal privilege that technological determinism grants to machines. At the same time, though, my analysis also undercuts the humanist inverse of technological determinism, which centers its explanations on the social rather than the technological. This point, too, must be clear from what has already been said, but it bears further discussion.

5.3 THE MANGLE, SOCIAL THEORY, AND LIMITS

Modern Americans confront a world in which everything changes, yet nothing moves.

David Noble, *America by Design*

The havoc interpretation wreaks in the domain of appearances is incalculable, and its privileged quest for hidden meanings may be profoundly mistaken.

Jean Baudrillard, "On Seduction," in *Selected Writings*

11. I can also mention a point of divergence from the actor-network analysis of technology and society, especially from the work of Latour. Latour notes that in general it is not easy to make facts and machines circulate outside the laboratory, and the history of N/C at GE certainly exemplifies this observation. But Latour (1983, 1987) also suggests that in order to ease such circulation the outside of the laboratory has to be made to resemble its interior. This is a very imaginative speculation, but the N/C story hardly bears it out. None of the transformations of the social at GE seem to have moved practices in the N/C shop toward those of the MIT Servomechanisms Laboratory from which N/C had

This section continues the confrontation between the mangle and traditional social theory that I began in section 2.5 and returned to in section 4.6. As it happens, David Noble offers his own traditional, nonemergent and humanist, sociological analysis of the historical episode that we have been examining, and in what follows I want to dissect the structure of that analysis and criticize it. My ultimate goal is to see how such analyses manage to sustain themselves in the face of historical evidence, both in a therapeutic spirit and as a way of further clarifying my own position.[12]

In the opening sentences of *Forces of Production,* Noble proclaims his humanism: "This is not a book about American technology but about American society. The focus here is upon things but the real concern is with people, with the social relations which bind and divide them . . . For this is the substrate from which all of our technology emerges . . . For some reason, this seemingly self-evident truth has been lost to modern Americans, who have come to believe instead that their technology shapes them rather than the other way around" (ix). It is clear from this passage, and especially from its concluding phrase, that Noble's distribution of agency differs from mine. He asserts the "seemingly self-evident truth" that agency resides in people *rather than* in the material world—people are the sole source of agency.[13] And this traditional and familiar move is followed in Noble's work by another one, equally traditional and familiar. Since human agency is the only agency around, it seems that one needs a positive account of it if one is to make any kind of sense of history, and Noble provides one, reaching for the classical repertoire of "interests." In his commentary on the story of N/C at GE, he focuses on management, to whom he imputes two enduring interests, an interest in making profits and an interest in dominating the

sprung. My conclusion is that the transformation of the social in technosocial mangling is a more open-ended business than Latour seems to allow.

12. Specifically, Noble's account is Marxist, but I am concerned with its overall form rather than its particular theoretical and political commitments. Having said that, it is worth noting that N/C has been an important topic for research and commentary ever since Harry Braverman (1974) singled it out as his principle exemplification of the Marxist deskilling thesis. Marxist scholarship has, then, served to constitute N/C as a key site for debate in social theory.

13. Making the same point, Noble (1979, vii) opens with a quotation from Mumford (1934): "[T]echnics . . . exists as an element in human culture and it promises well or ill as the social groups that exploit it promise well or ill. The machine itself makes no demands and holds out no promises: it is the human spirit that makes demands and keeps promises."

workers. And he further insists that the former is a minor consideration and that the latter—the interest in domination—is the really important one.

> It is a common confusion, especially on the part of those trained in or unduly influenced by formal economics (liberal and Marxist alike), that capitalism is a system of profit-motivated, efficient production. This is not true, nor has it ever been. If the drive to maximize profits, through private ownership and control over the process of production, has served historically as the primary means of capitalist development, it has never been the end of that development. The goal has always been domination . . . and the preservation of domination. (321–22)

Noble's humanism is thus of the usual temporally nonemergent kind, depending upon a stable set of interests—"the goal has always been"—to characterize a presumably stable set of actors: the dominators (capitalists/management) and the dominated (wage labor). Now, details apart, we all know this kind of story very well. The enduring interests of enduring actors, among which management and labor are prototypical examples, are stock categories for thinking about contemporary life in general. So what could be wrong with seeing N/C as just another example of this preformed scenario, as Noble suggests?

Actually, quite a lot. As a preliminary remark, I can observe that, even if one were to accept Noble's analysis, the *details* of the evolution of shop-floor culture would remain unaddressed. The only way to understand the specific transformations of the social in this episode is by reference to the particular resistances and accommodations that emerged in the course of it. This remark might, of course, be taken to point to the need for some eclectic conjunction of the mangle with traditional social theory, in which the former puts emergent, posthumanist flesh on a nonemergent, humanist skeleton. But I want to develop a critique that goes deeper than that, and that begins by noting that the inauguration and development of the Pilot Program at GE is surely a prima facie refutation of the idea that management's interest in domination (rather than profits) is truly fundamental to history. Labor gained more autonomy as its roles and relations shaded into those of management. Dominance was reduced, in this episode, not intensified.

The obvious commentary to give, then, if one wishes to think in terms of interests, is that the interest of management shifted during the history of N/C at GE. The interest in domination can be said to have intensified during the initial period of heightened Taylorism. The same interest might then be said to have declined as management implemented the Pilot Program, and then to have returned with renewed vigor on management's termination of the program in favor of a return to Taylorism. The interest of management, I feel inclined to say, should thus be understood as itself *within the plane of practice,* transformed and reconstituted there as part of the complex flow of resistances and accommodations already described. As an articulated self-understanding, at least, the interest of management was emergently and posthumanly mangled in this episode.[14]

And, it turns out, exactly the same commentary can be given about the workers. Thus the baseline for Noble's account is that labor has an enduring interest in wage rates as representing their share of the profits of industry (or perhaps they have an interest in protecting themselves from management's interest in reducing this share). However, it is clear that in the Pilot Program workers developed—emergently, in practice—what one could reasonably call in an interest in "the new way of life" (307, quoting a worker), an interest that was not specifically anchored in wage rates. "For some of them, as they later reflected, the Pilot Program was the most exciting thing they had ever been involved in, at work anyway, and they were loathe to give it up without at least a fight. As one pilot remembered, 'Some of the guys really didn't want to see it go. They were even willing to sacrifice the bonus—just don't bring back the foreman!'" (315). The workers' interests were, then, emergently and posthumanly mangled in this episode, just like those of management.[15]

14. Noble lets the subsidiary interest he imputes to management—in making profits—fade into the background of his analysis, but we can note that exactly what such an interest amounts to in practice is less straightforward and unambiguous than it might seem. Hopwood's fascinating archaeology of accounting systems (1987) shows that calculations of profit and loss can be embedded in a wide variety of record-keeping and computational systems, involving all sorts of assumptions and approximations, themselves transformable in practice in a way that I think should be understood as mangling. Discussing the economic rationale for N/C, Noble makes a similar point: "[Economic] justifications are most often made by the people who want to make the purchase, and if the item is desired enough by the right people, the justification will, in the end, reflect their interest. . . [P]ost-audits are rarely made, and when made usually are so designed as to ratify previous decisions . . . And the data is not all neatly tabulated and in a drawer somewhere. It is distributed among departments, with separate budgets, and the costs to one are the hidden costs to the others. In addition, there is every reason to believe that the data that does exist is self-serving information provided by each operating unit to insulate it from criticism and enhance its position within the firm. Finally, economic viability means different things to different people" (217).

15. Noble's account actually points to a wide range of interests at stake around N/C at GE. People in many job classifications feared for their livelihoods as the pilots en-

And, further, the traditional social-science assumption that this episode is well characterized in terms of a stable set of social actors is itself problematic. The defining property of the Pilot Program was, after all, that the role of labor blurred, to a certain extent at least, into that of management. There is a case, therefore, for saying that the very identities of the social actors involved were mangled—the precision of definitional boundaries that obtained before and after the Pilot Program was lost during it.[16]

On all of these counts concerning the stability of actors and their interests, then, Noble's analysis of his own historiography seems importantly misleading. But, it might be said, Noble is aware of this worry, and his analysis certainly includes a further stage that appears to take care of it. The key passage in *Forces of Production* reads as follows:

> If the actual requirements of production called for a relaxation of shop floor supervision and less authoritarian decision-making within the plant, such measures could be tolerated—but only *within limits.* In the final analysis, these *limits were determined* by a consideration far more fundamental than that of profitable production, namely, the preservation of class power.
>
> The Pilot Program *disclosed the limits,* for the workers and their unions, of such participation programs. The experience indicated that . . . to the extent that such programs actually serve as a vehicle for enlarging worker power, they will invariably *run up against the larger limits* set by capital and hence be terminated. (318, emphasis added)

---

croached upon their traditional functions; others were jealous of the pilots' autonomy and resented not being able to join them. Further: "Corporate managers were concerned: that they had not been adequately informed by Lynn about the program expansion; that the 10 percent bonus was disturbing the corporate-wide pay classification system; that the vertical enrichment notion raised the question of whether or not people could be fairly paid relative to other company positions; [and] that more fluid work group roles might not conform to National Labor Relations Board definitions of exempt (salaried management) and non-exempt (bargaining unit, hourly) jobs" (302). Higher management (like the union) thus, it seems, had an interest in corporationwide and industrywide standardization of working practices, pay rates, and so on, though Noble gives it no credit in his interpretive summary. All of these interests should, I think, be seen as products of the mangle, with individuals and groups formulating and reformulating them and their implications within the flow of practice. Miller and O'Leary (1994) document the diversity of groups, expertises, and interests that can bear on cultural extension in the workplace and the transformations that they can undergo within particular projects.

16. Zuboff (1987) discusses transformations that amount almost to losses of identity of labor and management in relation to the introduction of computer-based information technologies into the workplace.

"Limits" here are equivalent to the "constraints" that I discussed in section 2. 5. And their invocation enables Noble at once to admit a degree of mangling of actors and their properties while remaining within the traditional nonemergent humanist frame. Interests can indeed be mangled a little, but only within limits, which importantly continue to be definitively human and temporally enduring phenomena: the limits are defined in terms of "class power" and are always there—like fences or the walls of the prison, one "run[s] up against" them.[17] This is why "everything changes, yet nothing moves."

Noble's analysis, then, nicely exemplifies the class of positions in social theory that I contrasted with the mangle in section 2.5. The effects of the introduction of N/C at GE need to be understood in terms of stable actors and their almost stable interests that can only waver between enduring limits. As I said earlier, this form of analysis is familiar, and until I embarked on my own explorations of practice, I would probably have accepted it without thinking. It is flawed, though, at least as Noble articulates it. To see this, one has only to reflect upon the "limits" that are Noble's last ditch of defense against temporal emergence (and posthumanism). Two points are crucial.

First, although Noble might seem to be occupying a classically Durkheimian terrain, it is worth making it explicit that the limits he invokes are not, so to speak, *visible* ones. He cannot, for example, be referring to articulated work rules that publicly specify the roles and relations of production. On his own account, those were revisable in practice. The

17. In the passage quoted, Noble is talking about the interests of capital and management, but he handles the apparent mangling of labor interests mentioned above along the same lines, treating the emergent interest in the Pilot Program as, in effect, false or misguided, a mistake relative to some stable and unproblematic baseline. He concedes that "[p]articipation in such programs can indeed be a liberating and exhilarating experience . . . 'These people will never be the same again. They have seen that things can be different'" (318), which sounds like a recognition that interests can and do emerge and change along with specific projects. But then he makes his recuperative move: "the excitement and enthusiasm engendered by such programs, as well as the heightened sense of commitment to a common purpose, can also be used against the [presumably true and enduring] interests of the work force." There follows a long list of reasons why the workers' interest in the Pilot Program was mistaken, all predicated on the assumption of an ultimate reversion to traditional discipline, before Noble ties the package up neatly in a return to the "limits" at issue: "In short, the managerial contradictions inherent in the use of capital-intensive technology such as N/C, which prompted the introduction of the Pilot Program at GE, are embedded within the larger contradictions of capitalist production. And these contradictions place limits upon and ultimately chart the course of such participation programs" (319).

notion of "limit" that Noble needs is a behind-the-scenes one. But then, second, this notion of hidden limits is, as far as I can make out, both retrospective and idle. Its retrospective quality is evident in that the initial plausibility of Noble's formulation rests on his having already told the story of how the Pilot Program did, in fact, come to an end. The eventual termination of the program is what makes the idea that it ran into a limit possible. And the idleness of the appeal to limits can be shown by noting that *any* decision to terminate the Pilot Program at *any* point along its trajectory could trivially be glossed along such lines. If the program had been terminated sooner, one would simply assert that the limits on the flexibility of management's interest in domination were that much more restrictive. If the program had been allowed to go much further than it actually did toward erasing the boundary between labor and management before it was stopped, Noble could still write the same sentences. Even if it had never been terminated, the option would remain open to assert that "the limit" had not yet been reached. In the present instance, then, I think that limit talk marks "a kind of opaque endpoint to reflection and explanation," as Stephen Turner puts it (1994, 43); it points to the limits of Noble's analysis rather than to an important behind-the-scenes structure of practice.

The outcome of this discussion is, then, the following: Hidden limits (constraints, horizons) are, I believe, a necessary part of traditional social theory if the latter is to get to grips with—or protect itself from—the historical details of episodes like the one at issue, in which social actors and their properties were, on the face of it, posthumanly and emergently mangled. But limits so conceived are themselves deeply suspect. To be precise, appeal to them is vacuous unless their contours can be specified *in advance* of whatever episode is to be explained. And this is, at minimum, an exceedingly difficult requirement to meet. I find it hard to imagine, in fact, that there was any particular sticking point defined in advance as far as GE management and the Pilot Program were concerned.[18] My conclusion is, therefore, that, as Baudrillard puts it, the "privileged quest for hidden meanings may be profoundly mistaken,"

18. I had thought that I was more or less alone in my critical sensitivity to nonemergent prison metaphors in social theory until I came across the following passage: "The problem with the idea of objective limits [or constraints] on textual readings, or on descriptions of physical events, is that it is impossible to say in advance of discussion *what exactly they are,* outside of the circularity of taking the author's word for it, or appealing . . . to what other readers will find 'preposterous.' But that is unfortunate too, since what people in general find preposterous is patently a matter of social judgment and consensus,

and that we should prefer the mangle to traditional social theory. We should see the blocks to cultural extension as what they appear to be—resistances emerging in time—rather than the surfacing of otherwise hidden limits; and we need to recognize the existence of an impure, posthuman dynamics, reciprocally linking and transforming on the one hand the scale and boundaries of social actors, their social relations, disciplines, and goals, and on the other machines and their material performances. The alternative is the unreflective projection of commonsense categories and the consequent obliteration of fascinating social phenomena.[19]

---

and no more a guarantee of truth or reality than it is when later judgments declare everyone to have got it all wrong" (Edwards, Ashmore, and Potter 1994, p. 13 of draft).

19. It might be said that it is unfair of me to make this critique of traditional social theory hinge upon Noble's work. I would, for example, happily agree with the suggestion that the power and importance of Noble's writings derives from his historiography rather than the relatively few pages he devotes to explicit commentary and analysis. Nevertheless, it is relevant to note that Noble's subject matter, the industrial workplace, is the hard case for my critique. Like most people, as I said, my first impulse is to agree with Noble's analysis. All that I am doing in this chapter is assimilating the N/C episode to the examples of manglings of actors and their properties discussed in previous chapters, where countervailing intuitions to the mangle are much weaker if not nonexistent. And still, the moral is caveat emptor; the literature is full of appeals to limits and constraints that have no more substance than Noble's. For some examples of unreflective uses, see the quotations from Ginzburg, Latour, and Wise in chapter 2, note 36, but an anecdote will have to substitute for quantitative documentation. When I first became suspicious of limit and constraint talk, I started penciling small crosses in the margins of books and essays whenever I found taken-for-granted invocations of these concepts, thinking that I would publish a list one day. I gave up this practice when I discovered that almost everything I had read in the previous few months was tattooed with these marks. It is perhaps worth remarking that Noble's analysis is richer and more interesting than much of the literature at issue. The fear of human agency traditional to philosophy of science, and the Foucauldian reaction against human agency in the social sciences in general, both foster a tendency for authors to try to make do with only constraints ("method," in philosophy of science), suppressing the intentional structure of practice entirely.

PART TWO

Articulations

SIX

Living in the Material World

By attending only to knowledge as representation of nature, we wonder how we can ever escape from representations and hook-up with the world. . . Dewey has spoken sardonically of a spectator theory of knowledge that has obsessed Western philosophy. If we are mere spectators at the theatre of life, how shall we ever know, on grounds internal to the passing show, what is mere representation by the actors, and what is the real thing?

Ian Hacking, *Representing and Intervening*

With the four chapters of part 1, I have completed the task I set out in the opening chapter. I have exemplified and elaborated an analysis of practice in the performative idiom. I could therefore finish the book now. The details and summaries I have offered along the way just are, in embryo, my analysis of science (mathematics, technology, society). But no text stands in isolation, and in the two chapters of part 2, I want to locate my analysis in relation to other bodies of thought and work, in and around science studies. In this chapter, I let the tradition define the terrain. The central concern is with how the mangle bears upon our appreciation of scientific knowledge, and in particular how it bears upon the traditional philosophical problematics of realism and objectivity (rationality). Since these are the central topics of representationalism, they might seem to constitute a perverse point of departure for part 2. But they pertain to issues that are worth addressing, and their familiarity is important. They furnish some well-known landmarks from which to take our bearings and triangulate what is novel about the mangle, and I want to argue that the shift to the performative idiom can effect interesting and productive displacements in our thinking about them. This chapter also prepares some of the ground for the next, where the mangle, rather than the humanist/representationalist tradition, sets the agenda.

So, to put it schematically, the topography of this chapter is marked out by the three great narratives of traditional history and philosophy of science, the stories of Nature, Reason, and Society. The first of these is about the relation between scientific knowledge and the world, about the issue of realism and the correspondence (or lack of it) between representation and represented. In section 6.1, I present my own thoughts on realism and argue that the mangle points to a noncorrespondence realism (not antirealism) about scientific knowledge. In section 6.2, I emphasize what is distinctive about this kind of realism in a discussion of incommensurability, focusing on incommensurability in machinic performances. Section 6.3 then reviews some basic considerations that bear upon my discussion of the remaining two master narratives. The narrative of Reason is about a special rationality of science, usually conceived in terms of epistemic rules or standards of scientific method, that control (or should control) the production of scientific knowledge and thus guarantee its objectivity. The argument of section 6.4 is, first, that the mangle explicates a sense in which scientific knowledge can be said to be objective, but without any appeal to rules or standards; and, second, that whenever such standards figure importantly in scientific practice they should themselves be seen as subject to mangling. The narrative of Society pins scientific knowledge not to epistemic rules but to some property of the social, and hence points not to the objectivity of knowledge but to its relativity. In section 6.5, I argue that scientific knowledge should indeed be seen as culturally relative. But I seek to be clear on the nature of this relativism, and I argue that it cannot be specified in terms of substantive social or even technical variables (interests, gestalts, or whatever). Continuing this line of thought, the chapter concludes in section 6.6 by emphasizing that the temporal emergence of the mangle implies an irredeemably historicist understanding of science. To point up what is at issue, I incorporate here some additional material on Morpurgo's quark experiments and his controversy with another quark hunter, William Fairbank. The upshot of my analysis of practice in relation to these classical topics in science studies is, then, a noncorrespondence realism coupled with an insistence that, when these terms are appropriately construed, scientific knowledge is objective, relative, and truly historical, all at once.

6.1 REALISM

[Pragmatism] converts the absolutely empty notion of a static relation of "correspondence" . . . between our minds and reality, into that of a rich and active commerce (that anyone may follow in detail and understand) between particular thoughts of ours, and the great universe of other experiences in which they play their parts and have their uses.

William James, *Pragmatism*

What is the relation between our knowledge and the world? There is, of course, a well-developed debate on this topic in the philosophy of science that goes under the heading of "scientific realism," but this debate translates the question into a very limited and intransigent form. Traditional philosophy works within the representational idiom, and the space in which it can take on the realist problematic is exhausted by knowledge (empirical and, especially, theoretical) on the one side, and the world itself on the other. And just about the only philosophical question that can be constructed in this space is, does scientific knowledge mirror, correspond to, represent truly, how the world really is? To be more specific, as the debate has evolved in modern philosophy of science, the organizing question is, do unobservable theoretical entities appearing in theories that are empirically well confirmed correspond (perhaps in some approximate or convergent sense) to the state of nature? Are quarks, say, really the building blocks of the material world? or genes really the controllers of biological inheritance? Realists answer that, in some instances at least, such correspondence does obtain or, at least, that we are warranted in supposing so; antirealists deny that we have any such warrant (see Leplin 1984 for a range of contemporary positions).

In these debates, the realists have on their side our routine tendency to see knowledge as transparent, and our tendency to respect the hard work of well-funded communities of clever people. Who wants to say that all those particle physicists with their incredibly expensive machines and instruments have got it wrong about quarks? The antirealists, though, are in possession of a single, apparently unstoppable argument that can be elaborated in all sorts of ways. Barring mystical revelation or divine inspiration, science is generally regarded as being the best knowledge that we can have of how the world really is, so that it is impossible to imagine going behind the scenes of science to check whether science has, in this instance or that, got it right. Correspondence of scientific knowledge to the world is, then, ultimately uncheckable; assertions of correspondence are therefore nothing more or less than attestations of faith; and, say the antirealists, they should be disallowed as such.

Traditional arguments about scientific realism thus end, on the face of it, in a standoff, with realist intuitions permanently in tension with antirealist arguments. I will later, in fact, align myself with the antirealists in the debate over correspondence, but I want first to note that the analysis of scientific practice and the switch from a representational to a performative idiom can serve to displace the traditional problematic, to transform it into something rather different—more interesting, I think, and less intractable. Thus, in the performative idiom, one is not obliged to discuss the relation between knowledge and the world in terms of correspondence at all. One can ask instead about how, in practice, *connections between knowledge and the world are made, and of what those connections,* as made in practice, *consist.*[1] And, of course, chapters 2 to 4 above suggest answers to these questions.

I have argued that the connections between knowledge and the world are, in practice, interactive stabilizations of machinic performances and conceptual strata: these alignments of the material world with the world of representation are what sustain specific facts and theories and give them their precise form. And, if we are interested in the displaced problematic of realism, it helps to think about the temporal process of achieving such stabilizations. In chapters 2 and 3, we saw that captures and framings of material agency in machines and instruments depend on making a series of uncertain stabs along open-ended modelling vectors, and that the success or failure of those stabs cannot be reduced to anything in the human realm. Their outcomes depend on how the world is: does this machine or that instrument capture or frame material agency as intended? Further, in chapter 3, we saw that the framings of material agency that feed into empirical and theoretical knowledge constitutively and reciprocally intertwine the performance of machines and instruments with conceptual systems—we saw that a process of *mutual* mangling and tuning is at work there. Morpurgo had at once to struggle with the material performativity of his apparatus and with his interpre-

1. This point can be phrased as the complaint that traditional philosophy is too hasty and too lazy in its treatment of scientific realism. Instead of examining how the connections between knowledge and the world are made in practice, it ignores this process completely and reaches for one of the two answers that come immediately to anyone's mind—that knowledge does, or does not, correspond to the world. Star (1991b) calls this attitude "deleting the work," referring to the work of the scientists, not that of the philosophers, though the double entendre would be accurate. Gooding makes the point more positively: "The recovery of skills helps explain the apparent mystery of the successful referential function of much of scientists' talk; it is skilled agency that brings about the convergence of material and verbal practices. Convergence engenders belief in the correspondence of representations to things in the world" (1992, 104).

tive and phenomenal accounts of it, with no advance guarantee that any coherence of the three could be obtained. In this way, then, *how the material world is* leaks into and infects our representations of it in a nontrivial and consequential fashion. My analysis thus displays an intimate and responsive engagement between scientific knowledge and the material world that is integral to scientific practice. And, in this specific sense, the mangle thus offers us a realist appreciation of scientific knowledge. For the remainder of this section and the next, I want to make some observations about this mangle realism, and to explore some aspects of it.

My first point is that, while the realism of the mangle speaks of and specifies the relation between scientific knowledge and the world, the specification is different in kind from traditional philosophical realism (and its antirealist negation). The former concerns machinic performances and representational chains and how they are aligned with one another in time, while the latter dwells upon timeless relations of correspondence between knowledge and the world itself. To mark this difference, I call the realism of the mangle *pragmatic realism* (Pickering 1989c).[2] Second, I can note that pragmatic realism is, in the first instance, agnostic about correspondence; it is a noncorrespondence realism. Pragmatic realism specifies nontrivial links between knowledge and the world that are quite independent of relations of correspondence. Issues of correspondence, one can say, remain to be decided as far as pragmatic realism is concerned.[3] Third, pragmatic realism is a *nonsceptical* position, in Michael Lynch's terms (1992a). The key observation here is that the traditional realism debate gains much of its energy

2. Lenoir (1992, 166) also speaks of "pragmatic realism" in much the same sense. Barbara Herrnstein Smith has encouraged me to explain why I call my position "realist" at all, since its generative problematic is different from the usual one invoked by that word. There are two reasons. On the one hand, I cannot see why the philosophical tradition should be allowed a monopoly over the definition of this useful word. On the other, as a self-description of my position, it functions as a rhetorical roadblock to suggestions from philosophical realists that I am portraying science as a "mere construction" (Pickering 1990a).

3. "Pragmatic" thus signals both the reference to practice and an alignment with early pragmatist philosophy, especially with the work of William James ([1907, 1909] 1978), whose writings first suggested to me that one can explore how knowledge engages with the world without committing oneself to a position on correspondence. Dewey (1923, 308) remarks that "Peirce and James [and Dewey] are realists" in the sense of my pragmatic realism, while, as Hacking notes (1983, 62–63), the James-Dewey line in pragmatic thought ends up, as I do, being antirealist in relation to the traditional debate over correspondence.

from a characteristic humanist fear about representation, encapsulated in Hacking's reference to the "spectator theory" of knowledge.[4] The antirealist argument about the impossibility of going behind the scenes of science expresses a sceptical concern that representations, as human constructions, might never truly represent their object. There is no equivalent source of anxiety in pragmatic realism. The links between knowledge and the world are of the kind exhibited in my example of Morpurgo's quark searches, nontrivial achievements made in the real time of practice. There is no need to worry about them: we do not need to lie awake at night afraid that our knowledge has floated entirely free of its object.[5] In this sense, the pragmatic realism of the mangle defuses the entire problematic of correspondence realism; by supplying its own answer to the question of how scientific knowledge relates to the world, the mangle serves to stifle the impulse to ask about correspondence in the first place.

So the pragmatic realism of the mangle is an answer to the question of how knowledge relates to the world that differs in kind from more traditional answers concerning correspondence (either realist or anti-realist). And one important aspect of the move to the performative idiom is—in general, I believe—that it opens up a space for constructing new answers like this one to old questions. It makes available new and interesting things to think about. But I have not yet finished with realism. Although pragmatic and correspondence realisms come at the rela-

4. Hacking associates the spectator theory with John Dewey, but before Dewey, it was brilliantly derided by William James ([1907, 1909] 1978).

5. The same nonsceptical sentiment is implicit in the ethnomethodological concept of a "cultural object." Garfinkel, Lynch, and Livingston (1981) discuss the discovery of the first optical pulsar under this heading, and Lynch offers the following commentary: "In contrast to astrophysicists' usage, Garfinkel et al. speak of the IGP [the optical pulsar] as a 'cultural object' which is 'extracted' from a succession of observational runs with the optical and electronic equipment. They neither dispute nor adopt the claim that the IGP comes to stand as 'the cause of everything that is seen and said about it.' For the astronomers on the tape, the IGP becomes an astronomically specific articulation of the work through which they reflexively constitute it. Garfinkel et al. do not fetishize this product of the night's work; rather they insistently point to its genealogy in and as the astronomers' praxis. Nor do they discount its status as an object. It is not a representation, and it is no less of an object than any other (cultural) object" (1992a, 248–49). Lynch further remarks that "Garfinkel et al.'s refusal to speak of the optical pulsar as a trace on an oscillograph, a theoretical idea, a perceived figure, or an inscription divorced from any astrophysical object is not concomitant to a move into [correspondence] realism. It is more a matter of retaining an orientation to the 'chaining' of practical action and practical reasoning 'to the certain, technical, materially specific appearances of astronomy's *things*'" (249 n. 34).

tion between knowledge and the world at a different angles, still, as I shall now try to explain, the former undermines as well as defuses the latter.

If pragmatic realism is, in the first instance, agnostic about questions of correspondence, we can still ask how it might be mapped onto the rival positions in the traditional realism debate, and my suggestion is that in this mapping pragmatic realism comes out as a positive, non-sceptical, articulation/transformation of traditional antirealism. To put it the other way around, it is very difficult, to say the least, to see how the mangle can be aligned with traditional correspondence realism. One way to get at this point would be to note that my analysis of practice, as developed in part 1 and variously rehearsed in sections 6.4 to 6.6, points to a *situatedness* and *path dependence* of knowledge production. On the one hand, what counts as empirical or theoretical knowledge at any time is a function not just of how the world is but of the specific material-conceptual-disciplinary-social-etc. space in which knowledge production is situated. On the other hand, what counts as knowledge is not determined by the space in which it is produced. As I have emphasized, one needs also to take into account the contingencies of practice, the precise route that practice takes through that space. The contingent tentative fixing of modelling vectors, the contingent resistances that arise, the contingent formulation of strategies of accommodation, the contingent success or failure of these—all of these structure practice and its products. And these observations pose important problems for correspondence realism. To get to the latter from pragmatic realism, it seems necessary to suppose that the situatedness and path dependence of the mangle somehow (and sometimes) *wash out,* so that scientific knowledge eventually *converges* on a mirroring relation to nature, independently of where scientific practice starts from and whatever direction it sets off in.[6] I can only note that nothing in my analysis points to such an erasure of situatedness and path dependence, and this is the general sense in which the mangle weighs against correspondence realism. But this is to argue in the abstract, and I prefer to draw attention to a more specific sense in which the mangle creates problems for correspondence

6. Hence the idea of truth as the future consensus of a community of open-minded inquirers that has dogged pragmatist thought since its inception with Peirce. It seems to me that the following discussion of incommensurability deprives "future consensus" of the noncontingent associations of uniqueness that such ideas implicitly appeal to. I can imagine future consensuses on all sorts of incommensurable bodies of knowledge, but I cannot imagine how to tell which one of them corresponds to how the world really is.

realism. As Hacking notes (1983, 66), much of the agitation about scientific realism swirls around my next topic, incommensurability.

6.2 INCOMMENSURABILITY

Debates about incommensurability became central to the discourse of history and philosophy of science in the early 1960s, through the work of Thomas Kuhn (1962) and Paul Feyerabend (1962, 1965a, 1965b, 1975), and to explain what is at stake I can briefly rehearse Kuhn's position. Kuhn's idea is that the history of science consists of long periods of continuous growth—"normal science," governed by a "paradigm"—interspersed by discontinuities, or "revolutions." Further, Kuhn argues, in revolutions the world changes. Galileo and the scientists who came after him acted as if they lived in different worlds from their Aristotelian predecessors. Of course, just what one might mean by assertions like these needs clarification, and I will come back to this in a moment. But if one takes them at face value, some radical consequences follow.

If the world changes in revolutions, then the theories that apply to the world must themselves change. And, most importantly, there will be no neutral way to evaluate such theoretical transitions. Each theory will be appropriate to its own world, but not to those that precede or follow it. Theories that exist on either side of revolutionary discontinuities will be, as Kuhn and Feyerabend put it, incommensurable, meaning that they cannot be measured on the same scale. And at this point two intertwined sets of problems are posed for traditional philosophy of science. On the one hand, it appears to become impossible to defend traditional notions of the objectivity (or rationality) of science, since there is no way of adjudicating fairly between incommensurable theories. I return to objectivity in section 6.3. On the other hand, our realist intuitions are undermined. According to Kuhn and Feyerabend, revolutionary changes in science are typically accompanied by discontinuous shifts in the underlying ontologies attributed to the world by successor theories—the shift from phlogiston to oxygen at the birth of modern chemistry is an often-cited example. And if this is the case, why should we believe that scientists ever get the ontology right? Who knows whether the next scientific revolution might not just sweep away quarks or genes, just as earlier ontologies were swept away on the route to the present?

So we return to the realism debate with additional fuel for the antirealist: since Kuhn and Feyerabend, one standard antirealist move has been to cite incommensurability in the history of science as positive evidence against the idea that we have warrant for believing that any specified scientific theory mirrors how the world really is (see, for example, Laudan 1984a). This is a powerful argumentative strategy, but I do not intend to avail myself of it here. Instead, I want to run the argument the other way. I want to show how my analysis of practice speaks against correspondence realism by making incommensurability *thinkable in a new and straightforward way.*

The difficulty in getting to grips with the concept of incommensurability is that of imagining what "living in different worlds" might mean. How can one make sense of the assertion that in revolutions the world changes? How can Galileo and Aristotle have inhabited different worlds? Surely there is just one world that we all have in common, everywhere and always? Many people have taken the answers implicit in such rhetorical questions to require a rejection of the very possibility of incommensurability, but the point that I need to stress is that the entire argument between the opponents and defenders of incommensurability has, like the traditional realism debate more generally, largely remained within the representational idiom. The game has been taken to be one of showing how our ideas about the world can or cannot be said to influence our perceptions thereof. Speaking for the opposition, for example, Davidson (1973–74) argues in such terms that one can actually make no coherent sense of the notion of incommensurability, while, on the other side, Hacking (1983, chap. 5) discusses three different ways in which representational systems might be thought to infect our access to the empirical world, the most radical of which comes from one of Kuhn's original suggestions. Theory, Kuhn argues, functions as a social-psychological gestalt, a totalizing organization of our experience. And hence when theory changes, the world does, too—at least as far as "we" as theory-laden perceivers are concerned.

I do not want to enter into representationalist arguments for or against incommensurability, though they are interesting enough. I want instead to show how the move to the performative idiom leads one to expect "a new and fundamental type of incommensurability" (Hacking 1992a, 54) that relates to the distinctively machinic stratum of science. The point is this: On my analysis of practice, as I noted in section 3.4, the representational chains of science terminate not in "the world itself" but in specific captures and framings of material agency.[7] And there

7. Hacking thematizes the point thus, taking the discovery of the "Hall effect" as his example: "The apparatus was man made. The inventions were created. But, we tend to

seems to be no reason to suppose there exists any particular especially privileged—"true"—machinic mapping of the world. To the contrary, one can easily imagine that an indefinite variety of *machinic grips* on the world are attainable, attachable in all sorts of ways to all sorts of representational chains. Here, therefore, we glimpse the possibility of a form of incommensurability that cannot be articulated within the representational idiom, an incommensurability in captures and framings of material agency and in the representational chains that terminate in them.[8] This is the new way in which the mangle makes incommensurability thinkable, and to illustrate what I have in mind, I can offer an example.

In *Constructing Quarks,* I divided the history of elementary-particle physics into two phases, the "old physics" that was dominant until the early 1970s and the "new physics" that succeeded it, and I argued that the two were incommensurable. The theories and ontologies of the old and the new physics were clearly different from one another, but I centered my diagnosis of incommensurability rather on the disjuncture of the machinic bases of the two regimes, with their differing specific material performances. Thus I argued that the theories of the old physics were based on the phenomena most commonly found in the accelerator laboratory, while the new physics spoke instead of exceedingly rare phenomena, and I noted that each phenomenal domain was made available

feel, the phenomena revealed in the laboratory are part of God's handiwork, waiting to be discovered. . . I suggest, in contrast, that the Hall effect does not exist outside of certain kinds of apparatus. Its modern equivalent has become technology, reliably and routinely produced. The effect, at least in its pure state, can only be embodied in such devices" (1983, 225–26). Given that representational chains in science visibly terminate in machines and instruments, one might wonder how the idea that they actually terminate in "the world itself" is made available in science. The answer is back projection. The translation operators that convert material performances into facts, like Morpurgo's interpretive accounts of his apparatus, at the same time serve to *despecify* (or delocalize) those performances. Under the descriptions offered by any of Morpurgo's interpretive accounts, his specific material setups become instances of how the (electrostatic) world is *in general* (until the next destabilization, of course). Stabilized interpretive accounts, then, render particular material setups as transparent windows onto "the world itself." Higher levels of theoretical accounting pertain to "the world itself"—in general, rather than in particular settings—by deliberate construction, thus completing the vanishing act.

8. Papers presented at a workshop, "Materials in Science" (Princeton, 5 November 1993), foregrounded the extent to which biological knowledge has historically terminated upon particular species and materials—rats, mice, drosophila, cartilage, muscle. I believe one could develop a parallel argument about incommensurability there, too, though it might turn out to be an incommensurability of material rather than precisely machinic grips on the world. I am also reminded of Rudwick's history of geology (1985).

through its own battery of machines and instruments. At the most obvious level, the old physics was sustained by experiments that directed beams of particles at fixed targets, while the new physics was the physics of colliding beams. Where fixed-target experiment persisted into the new physics, the beams of interest became beams of leptons (electrons, neutrinos) rather than the beams of hadrons (pions, kaons) that had been the mainstay of the old physics. Further, the geometry of the laboratory was transformed in the move from the old to the new physics. In the former, detectors were positioned close to the axes of the beams to pick up the predominant "soft-scattering" events. In the latter, detectors were initially positioned perpendicularly to beam axes, to optimize their sensitivity to rare "hard-scattering" events. Later in the history of the new physics, targets were surrounded by detectors, and their output signals subjected to layers of computer processing to filter out soft-scattering events prior to physical analysis. And so on.[9]

In short, the transformation from the old to the new physics is an example of the kind of machinic incommensurability that the mangle leads us to expect. The theories of the old physics were tied to the world via the performances of one set of machines; the theories of the new physics engaged with a largely disjoint set. Representations of the world certainly changed in going from the old to the new physics, and they changed in an incommensurable fashion, since there was no neutral domain of machines, phenomena, and facts to which both could be held accountable. But no clever and debatable arguments about how representations infect the world are needed to make sense of this. The incommensurability resided in the shift of machinic grip made by the particle-physics community in the 1970s.[10]

9. Pickering 1984b gives all the details. Transformations in elementary-particle theory are discussed throughout the book; the gross transformations in experimental practice that accompanied them are documented in chapter 12; the discussion of incommensurability is in chapter 14. At a key point where the machinic bases of the old and the new physics coincided, interesting shifts took place in the interactive stabilization of captures and framings of material agency, interpretive and phenomenal accounts: see the discussions of the weak-neutral current discovery given by Galison (1983) and Pickering (1984b, secs. 6.4 and 6.5; 1984a; 1989b; the last of these includes a new discussion of the novel "cutting" procedures that were central to the discovery, and explores the divergence between my interpretation of this episode and Galison's).

10. For further examples and discussion of machinic incommensurability, see Ackermann 1985; and Hacking 1992a. More spectacularly, see Shapin and Schaffer's account (1985) of the argument between Thomas Hobbes and Robert Boyle about whether a machinic grip on nature should even be allowed to count in constructing knowledge. In one sense, the positions of Hobbes and Boyle were incommensurable, but the kind of

So where does this leave us? To conclude this section, four remarks are in order. First, as in the case of realism, the shift to the performative idiom effects an interesting and productive displacement of the problematic of incommensurability. Instead of thinking about how representations color our apprehensions of nature, it invites us to think about the machinic termini of our representational chains and how those termini can shift.[11] A very straightforward but nonetheless important sense of incommensurability thus becomes apparent. Second, this sense of incommensurability remains a nonsceptical one on the definition that I gave above. It does not require us to imagine that sometimes scientists get it right while at other times they are simply mistaken. The pragmatic realism of the mangle applies just as much to the old physics as to the new. There is no reason to doubt that the engagement of the old-physics representations with the world was just as sensitive and reciprocal as that of the new-physics. And thus, as mentioned earlier, we arrive at a *nonsceptical antirealism* (when seen in the terms of the traditional realism debate). Who knows how many different machinic grips on the world we can achieve (each with its own interactively stabilized facts, theories, and ontologies)? How would we ever know that we had achieved the "true" one? What might that even mean?

My third remark concerns the direction of my argument in this sec-

incommensurability at stake in the present section exists in a world where Boyle has already won.

11. To emphasize and perhaps to clarify this point further, I can note that discussions of incommensurability in the representational idiom naturally center on issues of language, and on the problems posed for transrevolutionary translation, communication, and understanding (see Kuhn 1970; and Davidson 1973–74). The kind of incommensurability presently under discussion has nothing especially to do with these topics: scientists working within the new physics had no difficulty in understanding what the old physics was about; nor did those physicists who remained faithful to the old physics have any deep problems in understanding the new. They just lived in different worlds, that was all. I can also observe that while Kuhn's concept of incommensurability in *The Structure of Scientific Revolutions* is a relatively heterogeneous one—embracing experimental techniques as well as worldviews and gestalts—his subsequent articulations have made the linguistic and conceptual aspects of incommensurability central (see, for example, Kuhn 1983). His 1990 presidential address to the Philosophy of Science Association (Kuhn 1991) was devoted to incommensurability as it might be understood within a particular linguistic theory of lexical taxonomies and structures, while in his latest publication that I have to hand he writes: "[W]hat makes these [scientific] specialties distinct, what keeps them apart and leaves the ground between them as apparently empty space? To that the answer is incommensurability, a growing *conceptual* disparity between the tools deployed in the two specialties. Once the two specialties have grown apart, that disparity makes it impossible for the practitioners of one to *communicate* fully with the practitioners of the other" (1992, 19–20, emphasis added).

tion. As promised, I have not taken the traditional antirealist route, arguing from incommensurability to antirealism. I have, in contrast, started from my analysis of the mangle (which I have developed without reference to incommensurability) and shown how it leads one to expect incommensurability of a certain form.[12] I am not committed, therefore, to any view of incommensurability as a primary feature of the history of science. I doubt, for example, whether Kuhn's schematic picture of periods of normal science interspersed with revolutionary discontinuities has general relevance. I am also not committed to the suggestion that incommensurability needs always be total, nor to the idea that revolutionary gulfs can never be filled.[13] My claim, rather, is to have foregrounded a possibility for dramatic change and difference in science that cannot be grasped in the representational idiom. This brings me to my fourth and final point.

In the move from the old to the new physics, much more changed than just knowledge and machines. Old skills and disciplines were devalued, and new ones became important. The bubble chamber, for example, found little use in the new physics, and a generation of bubble-chamber physicists had somehow to adapt to new ways of working around primarily electronic detectors. Social relations were also reconfigured, with new symbioses, as I called them, established in the circulation of research products between different traditions within the particle-physics community. Galison (1994) further notes that the growth of the new physics encompassed transformations of detailed technosocial relations extending outside the particle-physics community—to industry, for example, and, especially, the military-industrial complex.[14] In short, a whole slew of heterogeneous cultural elements—

12. I began the studies of practice that led me to the mangle partly because I was fed up with studying controversies and always concluding my interpretations by indicating their relation to discussions of incommensurability. I find that I have been led back to my old concerns by a new route. Fortunately for me, I have been led to other destinations as well.

13. I am happy enough to observe, for example, that at certain points—especially around the theoretical and experimental traditions concerned with the "constituent quark model"—the old and the new physics of elementary particles overlapped. Likewise, it does not trouble the preceding discussion to observe that attempts (not, in fact, strikingly successful) have been made to heal the disjunction between the theories of the old and new physics—as in computer simulations of lattice-gauge theories that seek to reproduce observed hadron spectra.

14. Thus, Galison asks rhetorically, "Is the evolution of supercomputer hardware and software for other purposes (cryptography, weapons design, meteorology, economic forecasting) internal or external [to computer simulations of lattice-gauge theory]?" (1994,

social as well as material and conceptual—was mangled and restabilized in the switch to the new physics. And this invites the question, why privilege the conceptual and representational strata in such transformations? Why, more precisely, have the potent notions of "incommensurability" and "different worlds" traditionally been irrevocably attached to discussions of difference at the level of theory? As usual, the villain of the piece is the representational idiom, and, as usual, the performative idiom suggests a way out of its claustrophobic space. The mangle, I suggest, encourages us to see *difference* in any of the cultural strata of science as interesting, and the intertwining of such differences as especially interesting, whether or not differences at the level of representation are involved. Here the mangle again points to a radical and productive displacement of traditional problematics.[15] In section 7.5, I take this line of thought further in some thoughts about difference and incommensurability in human (and nonhuman) powers.[16]

6.3 KNOWLEDGE AND US

Just as the problematics of realism and incommensurability are about the relation between knowledge and the world, so the problematics at issue in the remainder of this chapter are about the relation between

---

p. 8 of draft). He also notes that a "constraint" on the construction of certain kinds of new-physics "calorimeter" detectors was "the cost and availability of uranium within the whole . . . nuclear industrial and military cycle of production and consumption" (10), and further that "[a] young precision bubble-chamber experimenter, for example, must in the 1990s know not only about heavy quark decays but about the tough transparent material lexane—about the manufacturers that make it, about its optical qualities, its tensile, shear, and temperature-resistant strength, in short about the many properties that make lexane ideal for military canopies on jet fighters and which constituted its original reason for existence. A colliding beam experimenter has to learn not only about rare decays and Higgs searches, but about computer and electronic problems that are shared by electronics engineers preparing radiation hardening for war fighting on a nuclear battlefield" (13).

15. Such a displacement is already manifest in many of the works in cultural studies of science discussed in section 7.2; I make the point explicit in chapter 7, note 14.

16. Having drawn extensively upon Ian Hacking's work in this and section 6.1, I should say something about his preferred form of realism. Hacking (1983, chap. 16) argues for a kind of "entity realism." His idea is that we should be realist about entities like electrons, which scientists routinely use to intervene in the material world, though we should be less confident about theories. Electrons, as manipulated entities, can persist through many and drastic changes in the theories in which they are understood. The reference to intervention marks Hacking's shift to the performative idiom, which evidently I support, but it is worth noting that the pragmatic realism outlined in section 6.1 is, in the first instance, just as neutral about entity realism as about correspondence realism. Further, the discussion of machinic incommensurability in the present section would discour-

knowledge and us, about the degree to which scientific knowledge is distanced from specific human communities of producers and users (objectivism) or is irrevocably tied to them (relativism). Traditionally, objectivism and relativism have been defended and attacked as diametrically opposed positions, the one denying just what the other asserts. However, some of the most interesting recent thinking about science, and about knowledge more generally, has arrived at the conclusion that a simpleminded opposition between objectivity and relativity is untenable.[17] I share this conclusion, and my aim is to lay out my own route to it in terms of the mangle. The upshot is to show that the mangle delineates a position that is not so much "beyond objectivism and relativism" (Bernstein 1983) as objectivist, relativist, and historicist all at the same time.

The general form of the argument is simple enough, though I specialize and elaborate it in various ways in the following sections. If we can agree that practice as cultural extension is a process of open-ended modelling, then traditional objectivist and relativist accounts can be phrased and compared with the mangle as accounts of *closure*—of how particular modelling vectors get singled out from the indefinite range of possibilities. Thus objectivists insist that closure should be achieved relative to epistemic standards or rules—only extensions of theory that meet such standards deserve to be called objective (or rational).[18] The rules guarantee the objectivity of science by tying scientists' hands behind their backs in knowledge production. Relativists, instead, want to understand closure in one of two ways. Either distinctly social factors, like interests or social structure, determine the vectors of cultural extension, or something more internal to the technical culture of science, like a theory, worldview, or paradigm, functions to cut down the space for practice.

---

age me from making the move from pragmatic realism to entity realism. It is perhaps significant that Hacking's discussions of incommensurability appear in the "Representing" part of his book, and do not figure in the "Intervening" section. Hacking's more recent discussion (1992a) of incommensurability as extending into the machinic base of science makes no mention of entity realism.

17. For some examples of this line of thought, see Bernstein 1983; Cussins 1992a, 1992b; and the contributions to Megill 1991–92 and Porter 1992b, this last being concerned with scientific objectivity in particular, the others with objectivity more generally.

18. In the standard discourses of philosophy of science, objectivity and rationality tend to appear as two sides of the same coin: rationality pertains to the process whereby knowledge claims are evaluated, objectivity to the products of that process. Theory choice is, or should be, rational, and if it is, the theories chosen can be regarded as objective.

Several points follow. First, despite their undeniable differences, the objectivist and relativist positions are isomorphous in their insistence that something *substantive and enduring* (nonemergent)—rules, interests, worldviews, or whatever—is, or should be, responsible for closure. Second, the mangle is itself an analysis of closure—the preceding chapters are organized around precisely this topic—but the analysis of closure in terms of the mangle is different in kind from traditional ones. Instead of appealing to anything substantive and nonemergent to explain closure, the mangle points to temporally extended *processes*—to machinic, conceptual, and social maneuverings in fields of material and disciplinary agency, and to stabilizations and destabilizations of cultural elements and strata. A question thus arises concerning these two different ways of accounting for closure, substantive and processual, and in the remainder of this chapter, I want to argue that the mangle-ish account can both displace and subsume the traditional ones. The mangle, I claim, specifies a nonsceptical sense in which science can be said to be objective that is different from the traditional sense, but that includes a transformed version of it, and likewise a sense in which it can be said to be relative. At the same time, the temporally emergent character of the mangle gives rise to a historicist appreciation of science. And this is how, on my analysis, scientific knowledge can be simultaneously objective, relative, and truly historical.[19]

6.4 OBJECTIVITY

> Honda had secretly labeled his own fear the "sickness of objectivity." It was the ultimate hell, filled with pleasurable thrills, into which a cognition that refused to act was finally precipitated.
>
> Yukio Mishima, *The Temple of Dawn*

Epistemic rules offer one way of thinking about closure in scientific practice, the mangle a different way. In this section, I begin by establishing that the mangle's account of closure can genuinely be understood as an account of the objectivity of science. The mangle is, to this degree,

19. The following sections thus take up Husserl's idea that "[t]he point is not to secure objectivity but to understand it" (1970, 189). Lynch attributes a similar sentiment to Wittgenstein's *Philosophical Investigations:* "The point of the demonstration therefore was not to undermine objectivity, but to clarify 'in what sense mathematical [or scientific] knowledge can be said to be objective,' . . . which is not the same as arguing that such knowledge has an objective or transcendental foundation" (1992a, 226).

talking about something like traditional accounts of objectivity. Beyond that, I want to suggest that the mangle-ish account of objectivity subverts the traditional one.

To relate the mangle to objectivity, we can start with subjectivity. In part 1, I argued that one cannot understand scientific practice without reference to the intentional structure of human agency. Tentative vectors of cultural extension are fixed by human plans and goals, and, since I can find no general explanation of why specific scientists conceive specific plans, this aspect of my analysis points to a subjectivity of scientific practice: plans and goals are proper to their authors. The rest of my analysis, though, points away from such subjectivity. It emphasizes that initial plans do not determine outcomes. It emphasizes instead the redirection of modelling vectors in the course of dialectics of resistance and accommodation taking place in fields of material and disciplinary agency, dialectics that are themselves constitutive of the detailed products of practice. Further, these dialectics can entail the reformulation of plans that, anyway, collapse in machinic captures of material agency and in the achievement of interactive stabilizations of multiple material and conceptual cultural elements. The products of scientific practice, then, including knowledge (as well as machines, instruments, disciplines, social relations, and so on) are significantly detached by the mangle from the subjectivity of human agents (who are, nevertheless, integral to their production). This detachment from the intentional structure of human agency through encounters with material and disciplinary agency (themselves proper to no individual subject) is the basic sense of the objectivity of science that the mangle makes available to us.[20]

Now I want to draw attention to certain features of mangle objectivism, starting with the observation that this sense of objectivity is nontrivial. Passage through the mangle effectively defines a rather severe criterion of objectivity. This should be clear from the centrality of resistance to my analysis. The successful captures, framings, and interactive stabilizations that characterize the objective contents and prod-

20. Notice that this analysis of objectivity extends beyond the traditional locus of theoretical knowledge to the domains of facts, instruments, machines, and the social—though talk of objectivity seems increasingly strange as one proceeds along this list. Similar observations apply to the discussions of realism above and relativism and historicism below. Notice also that the dialectics of resistance and accommodation in engagement with disciplinary agency that I discussed in chapter 4 define a non-Platonist sense of the objectivity of mathematics. According to Tiles (1984), Gaston Bachelard held a similar view of objectivity as arising in the "interference" of different mathematical domains.

ucts of science are hard to come by; their achievement is difficult and uncertain. Most of the time scientists are submerged in a nonobjective mess. It is not, therefore, the case that "anything goes" in science in any sceptical sense. A mangle-ish analysis of objectivity thus means something; it is by no means idle.

My second observation is that, on my definition, scientific objectivity can be located already at the level of individual practice. I have, for example, described Glaser, Morpurgo, and Hamilton as successfully executing maneuvers in the field of material and disciplinary agency which, I believe, confer objectivity on the products of their practice prior to any social ratification. This way of speaking seems appropriate because the agencies that effect the delocalization of the products of individual practice are not themselves proper to any individual. But this is not to imply that objectivity is a once-and-for-all achievement of individual practice. As we saw in the Glaser and Morpurgo examples, interactive stabilizations achieved in practice can be once more destabilized in the subsequent practice of the same individual and, as the discussion in section 6.6 of Morpurgo's controversy with Fairbank will show, are likewise liable to destabilization in the practice of others. One might, therefore, want to set up a metric and say that items of scientific knowledge are more or less objective depending upon the extent to which they are threaded into the rest of scientific culture, socially stabilized over time, and so on. I can see nothing wrong with thinking in this way, but one should bear in mind that the process of communal ratification simply has the character of further manglings, and we have already analyzed it. To see this, we can just turn the telescope around and note that all of the passages of practice analyzed in part 1 were themselves part of processes of communal ratification—drawing upon existing culture, leaving some of it untouched, destabilizing and restabilizing other elements—at the same time as they were part of the production process for new cultural elements. And, following on from this, if there is nothing special and non-mangle-ish about this explicitly social dimension of objectivity, we have to recognize that, however highly it might score on our metric, stabilized knowledge is always liable to destabilization in future practice.[21]

So my claim is that the analysis of closure in cultural extension of-

21. Nickles (1992, 93–99) has an interesting discussion of this point under the heading of "single-pass versus multi-pass conceptions of science." Odd reflections upon normativity follow from the recognition that any knowledge is liable to be destabilized in

fered by the mangle can properly be described as delineating a sense in which scientific knowledge can be said to be objective. It is, though, a different sense from that traditional in philosophy of science. On my analysis, objectivity is a property of the products that temporally emerge from a posthumanly decentered process; traditional philosophy, instead, offers a nonemergent and humanist analysis. Objectivity there is seen as stemming from a peculiar kind of mental hygiene or policing of thought.[22] This police function relates specifically to theory choice in science, which, as I indicated in section 6.3, is usually discussed in terms of the rational rules or methods responsible for closure in theoretical debate.[23] The list of "grand methodologists" in the philosophy of science would include authors like Karl Popper ([1935] 1959), Imre Lakatos (1970), and Larry Laudan (1984b), and their proposed methodological rules for objectivity in theory choice take such forms as "propound only falsifiable theories," "avoid ad hoc modifications," "prefer theories which make successful surprising predictions over theories which explain only what is already known," and so on.[24] Such rules, it is im-

---

future practice. On the one hand, it seems proper to be impressed by a high score on the objectivity metric. On the other, to require that scientists somehow abide by the high-scoring elements of scientific culture would be to advocate a vicious conservatism that would retrospectively condemn the most impressive developments in the history of science: see Pickering 1991c.

22. "From Francis Bacon and Joseph Glanvill to Karl Popper and the Vienna Circle empiricists have wanted to police the methods of enquiry to ensure that science will be true to nature. That is the tradition in which I follow" (Cartwright 1989, 4). The police metaphor often turns into constraint talk, as when supposed epistemic rules are represented as constraints upon scientific practice: see, for example, McMullin 1988. It is in this guise that such rules come to figure in the "new eclecticism" in science studies (section 7.1).

23. The most action in recent methodological thought has centered on attempts like Allan Franklin's to extend the methodological approach to experiments by setting up a set of rules for their proper performance. Franklin thus seeks to extend classical discussions of objectivity to the empirical base of science (a topic hitherto neglected in the philosophical tradition but one that, of course, the mangle also directly addresses). For an argument between myself and Franklin on the same lines as that laid out below, see Franklin 1990, chap. 8; 1991; and Pickering 1991c; and for commentaries related to that debate, Ackermann 1991; and Lynch 1991a. Rasmussen 1993 is a wonderful story of the mangling of methodological rules (see below) in experimental practice.

24. I borrow the term "grand methodologist" from William Newton-Smith's talk "Science, Rationality, and Judgment" (1993). I thank him for a useful discussion on the present state of methodological thought in philosophy of science, and also for directing my attention to Laudan 1987 as a point of access to the recent state of play. The three methodological rules cited in the text are the first in a sample list of ten given by Laudan (1987, 23)—he is not recommending the items in this list; he cites them as representative

portant to note, have to be nonemergent—at least relatively enduring on the time scale of practice—if they are to accomplish their role in distancing the subjectivity of scientists from their product.

So, we have two articulations of objectivity before us: one following from the emergent and posthumanist analysis of practice that I call the mangle, the other consisting in some enduring humanist rules of mental hygiene, and for the rest of this section I want to inquire into the relation between them.[25] I will be as brief as possible because I suspect that few people remain interested in the methodological approach to objectivity, in part because it seems to have run out of steam, with those few philosophers still actively working in the area unable to agree among themselves what the master rules of science actually are.[26] Some significant points remain to be made from the perspective of the mangle, nevertheless. The form of my argument echoes that already given in respect of traditional social-science explanations in sections 2.5, 4.6, and 5.3.

My first remark is that in none of the case studies discussed here (or that I have discussed elsewhere) have I ever had an inkling that it might be useful or illuminating to invoke the kinds of rules that methodology talks about. This might, of course, point to some peculiar blind spot in the way that I address empirical materials and explanation, but it also points to a general conclusion. No finite set of rules can serve to explain practice and closure. The standard Wittgensteinian argument concerning the regress of rules for applying rules leads in that direction, but I prefer simply to observe that any explanation of the details of Glaser's, Morpurgo's, or Hamilton's practice clearly has at least to include reference to the specific dialectics of resistance and accommodation that

---

of the range of rules that have been recommended at one time or another (the first two are recognizably Popperian, the third Lakatosian).

25. Megill (1991) offers a typology and discussion of "four senses of objectivity": "absolute," "disciplinary," "dialectical," and "procedural." Unsurprisingly, the objectivity of the mangle falls primarily into the "dialectical" category, though it engages with the "disciplinary" and "procedural" senses, too. "Grand methodology" falls within Megill's "absolute" category. In what follows, however, I do not rest content with typological distinctions; I want to advocate my own analysis of objectivity against the traditional one.

26. "The theory of scientific methodology . . . appears to have fallen on hard times. Where methodology once enjoyed pride of place among philosophers of science, many are now skeptical about its prospects. . . Not everyone has given up on the methodological enterprise; but those who still see a prescriptive role for scientific methodology disagree about how to warrant that methodology. . . small wonder indeed that they cannot agree about which methodology to accept. All of which invites the perception that there is no good reason for anyone to pay the slightest attention to whatever advice the methodologist might be moved to offer" (Laudan 1987, 19).

structured their trajectories of cultural extension. Hence, to recycle a formulation from the discussion of Noble's social theory in section 5.3, a full appreciation of the objectivity of science might be taken to be an eclectic one—in which the objectivity of some broad features of science is explained in the humanist terms of nonemergent methodological rules, while the objectivity of specific details is accounted for in the emergent, posthumanist terms of the mangle. I am inclined, though, to a stronger conclusion, as I can now explain by a slightly roundabout route.

Philosophical rules of method are not supposed to be the ones stated by any particular scientist or by scientists at large. That would be too easy. They are instead tacit, and in need of philosophical explication and improvement; they lurk behind the scenes like the class interests that Noble talks about. From time to time, though, scientists do articulate rules intended to govern their own practice, and it is interesting to consider how this latter class of rules fits into my analysis. One instance springs to mind from the episodes discussed in earlier chapters. It concerns mathematics rather than science per se, but I think that the point I want to make has considerable generality anyway. In the mathematics of Hamilton's day, the so-called principle of permanence of forms was widely recognized. It was a rule that stated that no new operations should be introduced in the development of algebra that could not be given meaning in arithmetic. In effect, it defined what might count as a proper—rational, objective, or whatever—extension of the existing field of algebraic knowledge and technique, and what might not. And here is Hamilton's biographer on the greatness of Hamilton's mathematical achievement: "[Q]uaternions were the gateway to modern algebra. The Principle of the Permanence of Forms was shattered . . . and the road was open to a wide variety of algebras that did not follow the rules of ordinary arithmetic" (Hankins 1980, 301). Hankins is speaking here of the dependence of the quaternion system on noncommuting variables—variables for which the product xy does not equal yx—which have no equivalents in the arithmetical world of quantities and magnitudes. Noncommutation, which was quickly and enthusiastically put to use by many of Hamilton's contemporaries, was what "shattered" the principle of permanence of forms.

What interests me most about this little episode is that it indicates that explicit rules, at least, of rationality and objectivity like the principle of permanence of forms are themselves subject to mangling. In this instance, the principle stood in the way of the practice of Hamilton and others around quaternions—it constituted a resistance relative to various goals of subsequent elaboration of the quaternion system—and the accommodation of the mathematics community to this resistance was the direct and discontinuous one of jettisoning the principle. In the wake of Hamilton's work, it simply was no longer held to evaluate propriety, rationality, and objectivity in mathematics. This kind of explicit rule, then, needs to be seen as just one more element of scientific culture, and just as transformable in practice as the substantive elements of scientific culture that it "rules."

I can bring this line of thought into contact with the philosophical invocation of hidden rules by helping myself to an argument put forward by Stephen Turner (1994). What are we doing when we try to understand practice—and, by implication, objectivity—in terms of implicit rules? Turner's answer is that we must be reasoning by analogy; we must be thinking that implicit rules are like explicit ones. I think that Turner is right, and thus that the preceding discussion has rather general implications. My suggestion is that whenever we feel inclined to speak of implicit rules—this might occasionally be perspicuous—we should think of them as being like the principle of permanence of forms. Especially, I suggest, we should think of them as in the plane of practice and subject to mangling, instead of thinking of them as traditional philosophy of science is wont to do—as hovering nonemergently in some special epistemic heaven and controlling practice from without.[27]

The upshot is thus that methodological and mangle-ish articulations

27. Some further observations from Turner (1994) are also relevant. He notes that a typical difficulty of explanation by implicit rule is that the rules the analyst claims to discern (1) amount to little more than a redescription of whatever range of examples are originally chosen for examination, and that (2) the relevance of those rules outside the original domain of examples is contentious at best. This certainly seems like a good description of the situation in traditional philosophy of science, and it is just what one would expect if rules were mangled in practice rather than eternal universals thereof. I should note here that, while early attempts in philosophical methodology sought to spell out a fixed scientific method (Popper [1935] 1959), more recent attempts have conceded that the rules can change (slowly) with time. Laudan (1987), for example, argues that methods need to be seen as tailored to the aims of science, and that the latter have, in fact, changed historically. My analysis moves in the opposite direction from Laudan in this respect (as it moved in the opposite direction from Bloor in section 4.6). The "aims" that Laudan invokes are principles like "showing the hand of the Creator in the details of his creation" or "finding true theories about the world" (Laudan 1987, 22) and are supposed to exist on the far side of method from practice. My argument is that transformations of methods (and abstract aims too, I suppose, though I have never found it empirically useful to think about them either) need to be referred back to practice itself.

of scientific objectivity are not so much alternatives that one should choose between or eclectically combine. The analysis of practice subverts the former, displacing it onto its own terrain and suggesting that we see methodological rules, explicit and implicit, if and when they seem relevant to practice (and only then), as yet more mangle-able components of culture. The objectivity of the mangle, therefore, is prior to methodological objectivity and is always liable to subvert substantive articulations of the latter.[28]

6.5 RELATIVITY

The lion does not know subtleties and half solutions. He does not accept *sharing* as a basis for anything! He takes, he holds! He is not a Bolshevik or a Jew. You will never hear relativity from the lion. He wants the absolute. Life and death. No truces or arrangements, but the joy of the leap, the roar, the blood.

Thomas Pynchon, *Gravity's Rainbow*

In the academic world relativism is everywhere abominated.

Barry Barnes and David Bloor, "Relativism, Rationalism, and the Sociology of Knowledge"

Perhaps. And perhaps even beyond the academic world. Still, having discussed the objectivity of the mangle, I turn to the relativism that also follows from it.[29] A simple and straightforward relativist appreciation of science comes out of my analysis of practice. If existing culture is, as I have suggested, the surface of emergence for new culture, constituting the field of models and resources from which it is made, then knowledge

28. Bloor's discussion (1991, chap. 7) of formal and informal reasoning in mathematics and logic reaches the same conclusion. I can note that in discussing particular instances of historical development it is always possible to maintain an eclectic humanist/posthumanist sense that some enduring methodological rules were necessarily in play by retreating to a suitable level of abstraction in specifying those rules (and, of course, retrospection helps enormously). As more cases are considered, though, the effect is to tilt the balance of eclectic compromise farther and farther against such rules, eventually to the vanishing point.

29. For collections of essays for and, more usually, against relativism, see Wilson 1970; Hollis and Lukes 1982; and Krausz 1989. I will not go into the standard litany of complaints about relativism—that it conflicts with self-evident truths, that it is a self-defeating position, that it leads inexorably to epistemic, moral, political, etc., quietism, and so on. These complaints have been adequately addressed many times in science studies—see, for instance, Barnes and Bloor 1982; and Edwards, Ashmore, and Potter 1994. Beyond science studies, Smith (1988, esp. chap. 7; 1992; 1993b) rebuts them wonderfully.

is importantly relative to culture. This remark connects back to my discussion of incommensurability in section 6.2, and the contrasts between the old and the new physics of elementary particles can once more stand as an example of what is at stake. Each of these regimes drew upon its own more or less disjoint sets of models and resources—machines, instruments, theoretical and mathematical structures, and so forth—in elaborating its distinctive realm of facts, phenomena, and understandings of the world. Within each, then, knowledge as produced was relative to its specific cultural antecedents. This is the idea I gestured toward at the end of section 6.1, when I mentioned the *situatedness* of scientific knowledge. But the question is, just how is this cultural relativity to be spelled out? My answer, of course, is the mangle, and for the remainder of this chapter I want to inquire into just what that answer entails. Especially I want to emphasize that it entails that there are no general, substantive principles that we can hang onto in understanding cultural extension, that the connection between present and future culture is nothing more than whatever the temporally emergent trajectory of mangling turns out to be. I have argued this in the studies of part 1; all that remains is to make explicit the departure of the mangle from other, more familiar relativisms. The remainder of this section therefore takes on a rather negative character.

We can understand traditional relativisms in terms of the general schema introduced in section 6.3. The problematic can be rephrased as the question of what controls the extension of scientific culture; what determines closure in this process? Just as the traditional objectivist answers "rules," so traditional relativists supply their own nonemergent and humanist answers, answers that typically fall into one of two categories. The *social-relativist* response is that something distinctively social reliably structures knowledge production, while the *technical relativist* appeals to some element of the technical culture of science itself. I want to distinguish the mangle from these positions in turn.

Social relativism about science is associated with early formulations of the strong program in the SSK. There the idea was that some property of the social—social structure or social interest—endures through passages of practice, and that the technical components of scientific culture, including epistemic ones, are continually reconfigured around it.[30] The obstacles that the preceding chapters pose for this idea are obvious. I

30. To mention just one example, the first four chapters of Barnes 1982 offer a beautiful "finitist" exposition of the open-endedness of cultural extension, while the fifth and

have repeatedly argued that "the social" in all of its manifestations should be seen as just as liable to mangling in practice as the technical strata of science. The plans, goals, and interests of scientists; the scale and social relations of human actors; disciplines and forms of expertise—all of these, I have argued, are themselves at stake in practice; they do not control practice from without. The relativism of the mangle is not, then, social relativism. It does not privilege the social; the link between present and future scientific culture cannot be specified in such terms.

Now for technical relativism, where the same argument applies. The best-known version of this looks to the theoretical stratum of scientific culture to control the rest. The classic representational understandings of relativism (and incommensurability) are *top-down* ones, where high theory, worldviews, paradigms, or taxonomic lexicons necessarily structure all else—lower-level modelling; approximation schemes; instrumentation; the production, interpretation, and evaluation of data; and so on. Nothing in my analysis of practice supports this idea. All of the technical elements and strata of scientific culture are subject to the mangle, is my claim. If, as sometimes happens, entire constellations of practice become organized around a single theoretical structure—as in particle physics over the last twenty years—that, on my analysis, is a contingent upshot of practice. No deep principle lies behind it. Nor does any such principle lie behind the bottom-up empiricism of mainstream US sociology, nor any of the intermediates between top-down and bottom-up control that one can easily imagine and find examples of. On my analysis, there is only the mangle; and the mangle, I think, is just as capable of mangling such hierarchical arrangements as it is anything else. The temptation is, of course, to switch from a technical- to a social-relativist scheme to avoid this conclusion. One might want to think that the hierarchical control of, say, experiment by theory in science is a function of the relative social status of theoretical and experimental work—but that leads straight back into the argument against social relativism just laid out.

---

concluding chapter focuses on closure. Figure 5.2 (112) summarizes Barnes's analysis very clearly. It shows a horizontal arrow pointing from "old knowledge" to "new knowledge," while a vertical arrow points downward at the transformative nexus of an "accepted act of concept application." The upper end of this vertical arrow is labelled "goals and interests." In line with my earlier discussions of the non- or quasi-emergent property of the social in SSK, no arrow points *toward* "goals and interests" in this figure—on the page, goals and interests literally hover above cultural extension as they control it from without.

So, *the relativism of the mangle is not social relativism nor is it technical relativism.* It is a relativism to culture that cannot be specified in terms of enduring—nonemergent, nonmangled—properties of either the social or the technical in science. This is my basic point, which I want briefly to reinforce in a couple of ways. First, I emphasize that, while I claim that knowledge is relative to culture, what "culture" amounts to is itself subject to redefinition in practice, in at least two senses. On the one hand, to say that knowledge is relative to culture implies that the latter is somehow *bounded.* It is to assert, say, that there are distinct cultural fields to which the old and the new physics could be said to be relative. I think that this is right, but it is also right to notice that cultural boundaries can themselves be mangled, destabilized and restabilized, in practice. We came across an example of this in section 2.4, in the novel technosocial links that were constructed beyond the traditional domain of experimental particle physics in Alvarez's 72-inch bubble-chamber project. Material, conceptual, and human resources—in handling large quantities of liquid hydrogen, and from the developing world of electronic computers—that had hitherto lain outside the culture of particle physics were there incorporated into it. The scope of the cultural field to which knowledge is relative is itself, then, at stake in practice and not a given thereof (see Pickering 1989a for more on this line of thought). And, on the other hand, even if we have some notion of where the boundaries of culture are situated in some specific instance, still, the *contents* of that culture are not rigidly defined in advance of practice.

The key point here is that there is no explicit listing of the contents of culture that scientists can first consult and then seek to extend. Cultural extension as modelling is always conducted under some description, and there are indefinite possibilities for redescription in practice. Glaser's bubble chambers were modelled upon cloud chambers, but just how the latter should be described and extended went through several revisions en route to the former. Morpurgo's magnetic levitation electrometers were modelled upon Millikan's oil-drop apparatus, but that, too, was repeatedly reconceptualized as Morpurgo sought to develop it into something more useful for his own purposes. Hamilton played around with different elements of his mathematical culture, now treating this feature as intrinsic to algebra, now that. On the way to quaternions, he initially respected the two-part rule for multiplying complex numbers in the geometrical representation, for example, but then jettisoned the angle part of the rule as, in effect, irrelevant to a general description of multiplication. And so on. What is at stake here is the idea that the open-endedness of modelling itself amounts to an open-endedness in the description of culture. The culture to which knowledge production is relative is itself indefinitely reconfigurable. Where we are now is (temporarily) defined in getting to where we will be.[31]

To end this discussion of relativism, I want to comment on Peter Galison's "contexts and constraints" model of scientific practice (1987, 1988a, 1994; see also Baigrie 1994; and Pickering 1994a). As the reference to "constraint" suggests, this is a basically Durkheimian model, but Galison develops it in an interesting fashion. He emphasizes the importance of recognizing the multiplicity of scientific culture (and of the other cultures with which science engages). Cultural elements and strata act as contexts for one another, Galison argues, and constrain each others' development. And further, following Fernand Braudel, Galison suggests that layers of contextual constraints have differing inner temporalities.[32] Just as physical geography constitutes a very slowly varying constraint upon the evolution of human culture in general, so, Galison argues, there are "long-term" constraints upon scientific practice—"metaphysical commitments" of scientists, to principles like the conservation of energy, for example (1987, 246). And there are "short-term" constraints, too—for example, a scientist's "faith in a type of [experimental] device—even in a particular piece of apparatus" (1987, 254)—

31. Hence the impossibility that Kuhn (1970) notes of ever fully specifying what a paradigm (as "exemplar") is. There is nothing deep happening here. The cloud chamber, say, was a quite accessible object, no more shrouded in mystery than my cigarette lighter. But still, Glaser found many ways of describing it in the context of his particular project. I first grasped this point through Callon and Law's remark, which they also connect to Kuhn, that "*traditions should be seen as local constructions.* All translators . . . select and constitute traditions which will generally afford them a satisfactory environment" (1989, 79).

32. I think of Deleuze and Guattari's ideas about speed here, though these authors would not attempt a classification of temporalities and their ideas are more mangle-ish. "Pure relations of speed and slowness between particles imply movements of deterritorialization, just as pure affects imply an enterprise of desubjectification. Moreover, the plane of consistency does not preexist the movements of deterritorialization that unravel it, the lines of flight that draw it and cause it to rise to the surface, the becomings that compose it. The plane of organization is constantly working away at the plane of consistency, always trying to plug the lines of flight, stop or interrupt the movements of deterritorialization, weigh them down, restratify them, reconstitute forms and subjects in a dimension of depth. Conversely, the plane of consistency is constantly extricating itself from the plane of organization, causing particles to spin off the strata, scrambling forms by dint of speed or slowness, breaking down functions by means of assemblages or microassemblages" (Deleuze and Guattari 1987, 270).

in bubble chambers, say, or in Morpurgo's magnetic levitation electrometer.

So, in their own way, Galison's ideas about contexts and constraints make it possible to think of scientific knowledge as being both objective—in the sense that some temporal hierarchy of contexts exerts a kind of drag upon scientific practice independent of scientists' goals and plans—and relative to that selfsame hierarchy. At the same time, they indicate, as does the mangle, that it is possible to get by without specifying some single substantive link, like rules or interests, as the connective between present and future. I have stated my doubts about constraint talk already (sections 2.5 and 5.3). My examples speak against it. But Galison's constraints have the merit in this respect of being explicitly quasi-emergent at least—this is necessary to the idea that they have an inner temporality.[33] The question I want to raise, though, is whether we should follow Galison in thinking about cultural layers and strata as having such an orderly time profile—as coming into and out of existence on predictable time scales.

One point in favor of Galison's scheme is that it is certainly the case that not everything is transformed in each and every mangling. Some cultural elements and strata are just not "relevanced" in particular passages of practice. For example, neither Morpurgo's nor Hamilton's practice involved any interesting transformations of the social of the kind that we examined in chapters 2 and 5. As it happened, the trajectory of the mangle in the former instances left the social alone. At the same time, not even all of the cultural elements that are positively relevant to given passages of practice—that enter into intended relations of linkage or alignment—are necessarily mangled. Morpurgo, say, never tinkered with his pair of phenomenal accounts—they remained unrevised from the beginning to the end of his experimental program. Certain cultural elements and strata do, then, endure through particular passages of practice. And what Galison invites us to do is to wonder whether one can give some general account of why this is true of some elements in some situations and not others. In the end, I think one cannot.

33. Galison (1994, 2) says that "[t]he central historical-philosophical question . . . seems to me how these constraints arise, what sustains them, how do they act, and what makes them fall?" He exemplifies what he takes to be their action, but offers no suggestions on how they arise, persist, and fall. In this sense, Galison's concept of "constraint" is at most quasi-emergent, like Bloor's of "the social" (section 4.6). We can also note that Galison's examples of "constraint" are phrased in humanist terms: "metaphysical commitments," "faith in a type of device."

The problem is akin to that with Noble's "limits," namely, that Galison's classification of the temporal profiles of different cultural elements is a *retrospective* one. The only indication that some cultural element is a long-term rather than a short-term constraint, say, is that it has endured more or less unchanged for a long time. But without some additional input—like a dose of correspondence realism—this past endurance tells us nothing about what will happen tomorrow, about the likelihood that this or that long-term constraint will abruptly get mangled in the next passage of practice. Physicists had been constrained by their belief in the quantization of charge in units of the electron's charge for half a century until Gell-Mann and Zweig put forward their quark proposal—then they were not; particle-physics experimenters were constrained by the particular set of detectors available to them until Glaser invented the bubble chamber, and then they were constrained by a different set; international politics, to drift away from science for the sake of a telling example, were constrained by the adversarial relation between two enduring global superpowers, until one of them abruptly vanished in the late 1980s; and so on.[34]

My conclusion is, therefore, that while it is true that not every element and stratum of scientific culture is mangled in each passage of scientific practice, there is no especially informative pattern to be discovered about what changes and what does not. Galison's attribution of an innate temporality to cultural elements does not do the trick—at least if one's aim is real-time accounting rather than retrospective classification. We are stuck with the emergent posthumanism of the mangle. The links between the past, the present, and the future are specified by the mangle, nothing less and nothing more. In this sense, the relativism of the mangle is a hyperrelativism. It is on the wild side of the other relativ-

34. Thinking along similar lines, Popper remarks: "A trend (we may . . . take population growth as an example) which has persisted for hundreds or even thousands of years may change within a decade, or even more rapidly than that" ([1957] 1986, 115). One can make the point at issue very concrete. Law and Mol talk about "differences in durability" and remark: "The concrete bunkers still stand there, in the Dutch fields. Nearly fifty years of wet European winters have not undone all the work of those who built them. This is why concrete is so beloved of generals, architects, and road-builders. It often keeps on going. Is often durable" (1995, p. 4 of draft). Quite so—it has turned out that concrete is very durable. But the concern here is with practice as cultural extension, and who knows if the concrete will set and last in some *novel* situation (and who knows whether an apparently routine situation will *turn out* to be novel)? I wonder if concrete works in outer space—probably it has never been tried. Concrete (or "faith" in it) here is my analogue of Galison's contextual constraints—I say that concrete could get mangled tomorrow.

isms discussed in this section inasmuch as it offers us nothing substantive at all to hang onto in understanding cultural extension—not interests or social structure or gestalts or constraints of varying temporality or (to go back to section 6.4) epistemic rules or (section 4.6) metaphysics. At the same time, though, it is worth remembering that the manglings that constitute relativistic linkages through time are the selfsame manglings that confer upon science the objectivity discussed in section 6.4. The continual reconfigurations of the material, conceptual, and social strata of science that make it impossible to specify the relativity of scientific knowledge to any substantive variable are the upshot of the dialectics of resistance and accommodation that make the products of these transformations objective. The posthumanism of the mangle is, of course, what makes the trick possible. That is why there is no tension between objectivism and relativism when understood via an analysis of practice.

6.6 HISTORICITY

> In truth, the Legation knew, then, all that was to be known, and the true fault of education was to suspect too much.
>
> Henry Adams, *The Education of Henry Adams*

> Men . . . have discerned in history a plot, a rhythm, a predetermined pattern . . . I can see only one emergency following upon another.
>
> H. A. L. Fisher, *A History of Europe*

The mangle helps us grasp the sense in which science is both objective and culturally relative, but, as just discussed, it manages at the same time to locate cultural relativism somewhere in the ultraviolet of the traditional objectivist-relativist spectrum.[35] Traditional relativisms at least leave us something substantive to hang onto in thinking about cultural transformation, even if it is not the epistemic rules that traditional objectivism appeals to. And to mark this far-out position of the mangle, I am inclined to call it a *historicist* one. My usage of the term is, I think, fairly standard, but before I go any further, I should explain what I mean by it.

35. Callon and Latour note the similar impossibility of locating their actor-network approach in the space defined by traditional arguments about scientific knowledge; for a suggestive if somewhat enigmatic formulation, see their comparison of "SSK's yardstick" with the "'Paris' yardstick" (1992, 349, fig. 12.3).

In this book, I have sought to characterize what I take to be a regular pattern or structure of practice in terms of the concepts of modelling, resistance, accommodation, association, and so on. My claim is that this pattern repeats endlessly in scientific practice. At the same time, though, I have emphasized that brute contingency, sheer chance emerging in time, is integral to practice—in the tentative fixing of goals, in the emergence of specific resistances, in the substance of particular accommodations and their success or failure. These contingencies are irredeemably part of the structure of practice; they are constitutive of it rather than, say, interfering with it or being superimposed on it. And my conviction, arising from my studies, is that *they matter,* for our understanding of trajectories of cultural extension, and in our appreciation of the status of their products. Scientific culture, then, including scientific knowledge, has to be seen as intrinsically historical, in that its specific contents are a function of the temporally emergent contingencies of its production. This is what I mean by describing the mangle as sustaining a historicist vision of science. The historicism of the mangle, in other words, is temporal emergence seen from a different angle, looking backward instead of forward in time.[36]

To close this chapter, I want to try to bring home what is at stake in the historicism of the mangle by discussing, albeit briefly, a scientific *controversy.* This can help to highlight the consequentiality of contingency in a new way. I want to show that passages of practice starting from similar points and heading in similar directions can, contingently, lead to quite different outcomes. I want to foreground the *path dependence* of knowledge production, as I called it in section 6.1. It is also relevant here that scientific controversy has always been the preferred site of argument for SSK. I have indicated points of divergence between the mangle and SSK throughout the book, most recently in section 6.5, but discussion of a controversy will make it possible to formulate this divergence on SSK's home ground.[37] The controversy I have in mind is

36. As a benchmark, it might be useful to situate myself relative to Popper's critique of, as his title puts it, *The Poverty of Historicism* ([1957] 1986). The position that Popper attacks as "historicism" is a form of Marxist teleology that posits the existence of laws of social evolution—laws that require that society necessarily progresses through a fixed series of social forms, eventually arriving at socialism itself. The idea that cultural extension unfolds under some such law is the *antithesis* of my arguments about temporal emergence; the "historicism" that Popper attacks is almost the inverse of the historicism I ascribe here to the mangle.

37. I thank Trevor Pinch for pressing me to explain how controversy fits into my analysis of practice.

one that embroiled Morpurgo and his quark-search experiments in the second half of the 1970s, so what follows continues the narrative of chapter 3.

Morpurgo was not the only physicist who set out in the mid–1960s to look for isolated quarks in bench-top experiments. Only one other physicist, however, pursued them with comparable intensity, William Fairbank at Stanford University. Under Fairbank's guidance, a series of Stanford graduate students developed and performed their own series of Millikan-type quark searches. Fairbank and Morpurgo, then, had much in common in terms of resources and goals. Their trajectories nevertheless diverged. Exemplifying once again the open-endedness of modelling, the material procedures developed at Stanford were similar but not identical to those in Genoa. Like Morpurgo, Fairbank adopted a diamagnetic levitation system to float samples between two metal plates used to apply an electric field. But there the two went different ways. Superconducting materials are perfect diamagnets, and, as an expert in the low-temperature techniques required to sustain the phenomenon of superconductivity, Fairbank immediately thought in terms of a material procedure based on superconducting niobium balls. There were other differences of detail between the material procedures developed at Genoa and Stanford, but the respective choices of room-temperature levitation of graphite and low-temperature levitation of niobium were the most conspicuous.[38] And, although they were slow to produce results, from the mid–1970s onward Fairbank and his students published a series of reports on their experiments, claiming to have found positive evidence for the existence of quarks (for details, see Pickering 1981b). Through the second half of the 1970s, therefore, the situation obtained in physics that two very similar material procedures, each modelled on Millikan's and understood in terms of very similar interpretive accounts, sustained diametrically opposed accounts of the material world: quarks, for Fairbank, no quarks for Morpurgo.

What should we make of this divergence between Morpurgo's and Fairbank's findings? In early SSK, the existence of disputes and controversies within science served as an entering wedge for social-relativist arguments. The argument was simple, and went along the lines dis-

38. Professor Morpurgo emphasized to me in conversation (May 1993) that an important correlate of the choice between room-temperature and low-temperature techniques was that, with the former, one can actually see the levitating sample while, with the latter, one can only register its movements electronically (using SQUID detectors).

cussed in section 6.3 above. In controversies, the participants themselves make clear how hard it is to see knowledge as being given by the world, alone or in conjunction with some set of rational rules or method. And so, the argument went, something else must be invoked to explain why the controversial positions in question get taken up— namely, the social. Features of the social—like interests or constraints—thus fill the gap created by the inadequacy of the world and reason to specify knowledge and belief.[39] We have, in fact, already seen this argumentative strategy exemplified in David Bloor's discussion of mathematics (section 4.6), though there it was addressed to Hamilton's metaphysics rather than to the technical content of his mathematics. It also organizes the entire structure of Shapin's authoritative review of the empirical literature on the SSK, a review that describes my history of the Morpurgo-Fairbank controversy as opening the way to, though not actually accomplishing, a full-blooded sociological explanation (1982, 163). And Barry Barnes is less circumspect than Shapin, citing my study as already explaining the divergence between Morpurgo and Fairbank in terms of "vested interests" and "micro-political considerations" (1982, 115).

I mention this argumentative strategy of SSK because the mangle makes it, at least, redundant. As far as I have been able to understand the dispute between Morpurgo and Fairbank, and others like it, no SSK-like analysis is called for. On my analysis, no substantive variable fills the gap between the world and our knowledge—there is only history and the mangle. In chapter 3, I traced out Morpurgo's route to his findings in terms of the particular vectors of cultural extension that he pursued, the particular resistances and accommodations thus precipitated, and the particular interactive stabilizations that he achieved. The same could be done, I am sure, in respect of Fairbank. And these tracings are all that needs to be said about their divergence. It just happened that the contingencies of resistance and accommodation worked out differently in the two instances.[40] Differences like these are, I think, continually

39. Fine (forthcoming) gives a nice analysis of this chain of reasoning. I should make it clear that it is not peculiar to the non- or quasi-emergent articulations of SSK mentioned next. The pioneer of the controversy study was Harry Collins, whose writings often point, as noted already, to fully emergent phenomena. Still, for Collins too, the social fills the gap: see Collins 1992.

40. It is, of course, interesting to track the Morpurgo-Fairbank controversy further, but it does not change the upshot of this discussion. The differing results of the two groups constituted, in an obvious sense, a resistance to the construction of a consensual account of the world, and various strategies of accommodation were evident (all documented in Pickering 1981b). Many physicists, for example, simply assumed that the Stanford results

bubbling up in practice, without any special causes behind them other than the hazards of practice. On my analysis, the present is always incipiently and emergently breaking up, branching endlessly into the future—with no need of "the social" or anything else to prod it.[41] The trajectory of cultural extension is, therefore, truly historical. And hence the historicism of the mangle—which, I insist, sits alongside, not in opposition to, its sense of the relativity and objectivity of science.

---

were in error (since these, rather than those from Genoa, were in conflict with prior beliefs and findings). The Stanford experimenters, however, sought to refine their material procedures and interpretive accounts in the face of various critiques—replacing the Pyrex plates used to apply the electric field to the samples with optically flat and parallel quartz ones, and introducing a small copper ball between the plates in order to measure, and thus obviate theoretical estimation of, any fixed dipole moment on the niobium balls (Pickering 1981b, 226)—and continued to find evidence for fractional charges. An alternative strategy of accommodation consisted in theoretical attempts to reconcile the Stanford and Genoa findings. Interestingly, these attempts were at the level of quark theory, and amounted to revisions in the phenomenal accounts that Morpurgo and Fairbank had hitherto held constant. They also suggested new strategies for looking for quarks—especially as attached to heavy elements—but experiments designed along those lines failed to find any new evidence, and the suspicion remained that Fairbank's results were due to some unlocatable source of error (Pickering 1981b, 230–33). And so on. One can connect the history of these developments back to my remarks in section 6.4 on a metric of objectivity, as measuring the fluctuating extent to which the Stanford findings were woven into the technical practice of others.

41. I think Lyotard (1988) makes the same point, but in a representational rather than a performative idiom.

SEVEN

Through the Mangle

The grand narrative has lost its credibility.
Jean-François Lyotard, *The Postmodern Condition*

If this book ended with chapter 5, several new ones could start here. I content myself with touching a few bases. My aim is to position the mangle in relation to other nontraditional approaches to science studies and to indicate some directions for further thought and research. I address two overlapping concerns. One is my conviction that a new synthesis is under way in science studies. In section 7.1, I begin to characterize it, contrasting two integrative approaches that I call multidisciplinary eclecticism and antidisciplinary synthesis, and associating the mangle with the latter. Mapping the new synthesis in more detail (but still partially), in section 7.2, I discuss a cultural-studies approach to science. I argue that this approach and the mangle are orthogonal but complementary components of the new synthesis, and toward the end I relate cultural studies of science to the field of cultural studies more broadly understood.

My second concern in this chapter is with the relation between the micro and the macro in the analysis of practice. The focus of the examples discussed in part 1 was fine-grained, and usually on the practice of individuals or small groups (although the discussion of numerically controlled machine tools in chapter 5 opened out onto wider horizons). In sections 7.3 and 7.4, I therefore seek to explore how my analysis might feed into concerns with STS (science, technology, and society) in the large. Having already had my say on topics in philosophy and social theory, I concentrate on historiography, where I think the mangle can lead us in interesting directions that have yet to be mentioned. Section 7.3 discusses a shift in historiographic sensibilities induced by the move from the representational to the performative idiom, and argues for a performative history of STS. Section 7.4 exemplifies the possibility of conceptualizing such a history as macromangling in a brief discussion of the intersection of science and the military in World War II.

The chapter concludes with two afterthoughts. Section 7.5 discusses nonstandard human and nonhuman agency—agency that evades the characterizations that I have taken for granted in the rest of the book—and an associated conception of an incommensurability of powers. Section 7.6 ventures into "theory." I play with the idea of taking the mangle seriously as a TOE, a theory of, literally, everything.

7.1 ANTIDISCIPLINE: A NEW SYNTHESIS

History and Philosophy of Science: Intimate Relationship or Marriage of Convenience?

Ronald Giere, title of a 1973 article

[T]he distance between mainstream history and philosophy of science is probably greater now [1989] than it has ever been.

Larry Laudan, "Thoughts on HPS: 20 Years Later"

It has long been noted, and often bewailed, that science studies is a fractured field. In general, historians, philosophers, and sociologists of science have not got on well. When they have not gone their separate ways, the result has been border wars. Much the same can be said of relations between students of science proper, of mathematics, and of technology, and of the relations between the science-studies microdisciplines en masse and the parent disciplines—general history, philosophy, sociology. There a mutual disdain typically rules. This situation is changing, though, in a way that is worth making explicit.

As far as the traditional science-studies disciplines are concerned, the prospects for synthesis are not rosy. The abysses that yawn between and around them are built into their basic conceptualizations of the science object. Philosophical analyses of science built around epistemic rules (even if softened as "naturalized epistemologies") or sociological analyses built around social interests (or social structure) are just *different* theories of science.[1] The farthest that one can go with them toward syn-

1. Though I cannot take this line of thought further here, it is worth noting that specific conceptualizations of science serve to reinforce traditional disciplinary identities, and that this tends to hold them in place (Pickering 1993a). Thus, for example, analyses that revolve around epistemic rules immediately announce their allegiance to philosophy as an

thesis is to add them up. One can try imagining that science is ruled by rules *plus* interests, where the "plus" modifies neither of the terms it conjoins, but this does not erase the fractures. Syntheses within the tradition can, then, amount at most to an *eclectic multidisciplinarity.*[2] Giere conveys the flavor of this approach.

> I conclude that a combination of naturalistic realism, interest theory, and a systems approach provides at least the beginning of an enlightened postmodern synthesis. This suggestion proposes that the ideal treatment of any scientific or technological development is not one that analyzes it from a single, unified perspective, but one that successfully integrates diverse component perspectives . . . So the goal is not "unification" within a single perspective but the "integration" of several different perspectives. Rather than *inter*disciplinary research, one should think in terms of *multi*disciplinary research. But this probably requires more collaborative research than is the current norm in science and technology studies. (1993, 108)

This kind of synthesis thus amounts to a continuation of traditional divisions of labor, while its novelty lies in its recommendation of a summation of disciplinary products. It proposes a marriage of convenience, one might say. There is, though, a growing body of nontraditional work

autonomous field of practice. For a collection of interesting essays on disciplinarity and knowledge production, see Messer-Davidow, Shumway, and Sylvan 1993.

2. Eclectic multidisciplinarity has come increasingly into vogue in science studies since the mid–1980s: some examples are Campbell 1993; Fuller 1992; Giere 1988; and Hull 1988. On the history of eclecticism and its relation to the "internal/external" debate in science studies, see Shapin 1991. Eclecticism precipitates what I call the balance problem, the problem of specifying which of the multiple set of epistemic and social factors is dominant and when—a problem usually resolved in favor of the epistemic "in the last instance." In Pickering 1991b, I show how this works out in Giere 1988, and in Pickering 1990a, I show that the balance problem does not arise in the mangle. The eclectic incorporation of social theories into hitherto socially insensitive frameworks is not unique to science studies. I am reminded of Adrian Wilson's discussion of the history of social history: "What I call the 'social-history paradigm' consists of the application to history of concepts, methods or inspiration taken from one or another of the social sciences... The social sciences would bring about a 'refinement of the historian's social vocabulary,' statistical methods would confer precision upon historians' statements, and the disciplines of social anthropology and social psychology would provide the historian with new and incisive questions. This new conceptual apparatus ... would permit historians at long last to explain, as political and constitutional history had not, 'the workings of human society and the fluctuations in human affairs'" (1993, 15–16).

in science studies that displays the possibility of an intimate relationship, a more thoroughgoing synthesis. This is a body of work that is variously sensitive to science's performative, posthuman, and historicist aspects, and which, I believe, genuinely fits together to construct a common vision of science. Given the organization of the university, such work of necessity emanates from departments of philosophy, history, sociology, anthropology, and so on. But it is, at most, inflected by disciplinarity, rather than wearing disciplinarity on its sleeve in its a priori conceptualizations of what science is and how it should be studied. This work is antidisciplinary, at least when seen from the standpoint of traditional disciplinary conceptualizations, and in sum it either promises or already constitutes an *antidisciplinary new synthesis,* in which the microdisciplinary fractures in and around science studies are more or less erased (Pickering 1993a).

Such, at any rate, is my conviction, and my hope is that the preceding chapters, as well as what follows, will be read as a contribution to that synthesis. I will turn to its substance in a moment, but first I should observe that there is some potential for confusion between the mangle and multidisciplinary eclecticism. From my perspective, the eclectic recognition of the importance of both the epistemic and the social in science is certainly a step in the right direction, a step toward an appreciation of cultural multiplicity in science, and a fruitful departure from traditional monodisciplinary images of science. The mangle is, however, just as corrosive of the eclectic position as it is of its constituent parts. The point to note is that, like its disciplinary components, the eclectic analysis of science privileges certain traditional variables—epistemic rules, social interests—as enduring explanatory causes of scientific practice. The mangle, in contrast, strips away that privilege. I have emphasized that such variables are often not illuminating at all in the analysis of practice, and that, even when they are illuminating, they should be seen as in the plane of practice, just as subject to transformation there as conceptual structures, disciplines, social relations, and what have you—just as much at stake in the construction and destruction of heterogeneous interactive stabilizations as the rest of scientific culture. The mangle thus deflects our attention from any special concern with the particular variables that the disciplines traditionally invite us to focus on and conceptualize in a peculiar way, and directs us instead toward the unitary terrain of practice in a space of indefinite cultural multiplicity. Far from suggesting or reinforcing a position of eclectic multidisciplinarity, therefore, the mangle thus provides *a rationale* for antidisciplinary synthesis.[3]

7.2 CULTURAL STUDIES AND THE MANGLE

Cultural studies is part of the noise made by the great academic ice-floes of Literature, Sociology, Anthropology and so on, as their mass shifts and breaks apart.

Stefan Collini, "Escape from DWEMsville"

Now for the substance of the new synthesis. In actively drawing upon particular nontraditional "philosophical," "historical," and "sociological" approaches in science studies as this book has gone along, I have implicitly laid out my own (partial and incomplete) map of the new synthesis and of how the various bits and pieces might fit together. I could, therefore, just reproduce here a list of earlier citations as an extensive definition of how I see different approaches to science studies merging with one another.[4] More ambitiously, I could attempt a long interpretive and reinterpretive literature review along the lines of Shapin 1982, an essay that did much to give SSK substance in the 1980s. But the former approach would add little, and the latter is beyond the compass of this book. Instead I will adopt an intermediate strategy that might be perspicuous and that can both help to foreground the novelty of the mangle from another angle and, I hope, point to fruitful alliances

3. Actor-network theory can be understood as arriving at the same destination, though this has never been clearly spelled out. I think that antidisciplinarity is what is being recommended in Callon and Latour's frequent attacks on "sociology" and the "social sciences." Hence, when Latour writes that "the social sciences are part of the problem," and that "we strongly reject the helping hands offered us by the social sciences" (1988a, 161, 165), we should read him as reacting against traditional nonemergent and humanist circumscriptions of the methods and subject matter of sociology as an academic discipline. Callon is more explicit: "To transform academic sociology into a sociology capable of following technology throughout its elaboration means recognizing that its proper object of study is neither society itself nor so-called social relationships but the very actor networks that simultaneously give rise to society and to technology... This notion makes it possible to abandon the constricting framework of sociological analysis with its pre-established social categories and its rigid social/natural divide" (1987, 99–100).

4. It would be disingenuous of me not to note the frequency with which I have drawn on and cited the essays collected in the volume I edited, *Science as Practice and Culture* (1992). As I argued in the introduction to that volume, I think that the integrity of this collection of papers by "philosophers," "historians," "sociologists," and "anthropologists" of science instantiates an antidisciplinary synthesis.

within and beyond science studies. I will discuss the approach to science studies that, harking back to the definitions at the beginning of chapter 1, I call cultural studies of science. Such studies are, I think, orthogonal but complementary to the mangle, and, by defining a second axis, they help to delineate the space of the new synthesis. At the end of the section, I discuss the relation between cultural studies of science and cultural studies per se.[5]

In section 1.1, I made a distinction between "practice" and "culture." I offered a crude but broad definition of "culture" as "the 'made things' of science, in which I include skills and social relations, machines and instruments, as well as scientific facts and theories," to which I could now add "and so on," and I stated that the problematic of this book was to get to grips not so much with "culture" as with "practice," understood as the work of real-time cultural extension and transformation. I also remarked that the past decade has seen something of a bifurcation in science studies. My interest in the temporality of cultural extension is hardly definitive of the field. Instead, it appears to me that some of the most interesting and important recent work has concentrated on synchronous mappings and explorations of scientific culture, either through historical and ethnographic investigations of particular cultures at particular times, or through social and philosophical inquiries into the nature of scientific culture more generally. This, then, is the body of work that I call cultural studies of science, and its orthogonality to studies of scientific practice is evident: the one is about what scientific culture *is* at some given place and time, the other is about how culture *changes* in time.[6] One aim of what follows, however, is to establish that orthogonality does not imply contradiction and that, in fact, cultural studies of science and studies of scientific practice reinforce one another in elaborating, albeit in different ways, a shared vision of science.

5. I thank Claudine Cohen, John Law, Joseph Rouse, and Simon Schaffer for enlightening conversations and communications on earlier drafts of this section.

6. Simon Schaffer emphasized the orthogonality of the two approaches to me. He notes: "Much recent historical inquiry has been devoted to the labor processes in [the workplaces of physics]. Two important changes have accompanied these inquiries. First, attention to texts has been displaced by attention to practices... Second, histories have been displaced by maps. Chroniclers of the sciences often assumed that their task was to tell stories about the temporal development of natural knowledge. Contemporary science studies, however, appeal at least as much to the metaphors of geography... Maps of science and networks of skills and techniques have become commonplace in recent work... The power of physics over its world requires an extended and diverse imperium of complex practices" (1993b, 1).

Two further introductory remarks. First, as far as I can make out, the split that has developed between studies of culture and practice is a de facto one, unrelated to any differences of principle. It just so happened that many scholars found fascinating and productive avenues into the exploration of cultures while I remained obsessed with cultural transformation. Second, I should acknowledge that my work on the latter has made me particularly sensitive to issues of temporality. I doubt whether the relative atemporality of what I am calling cultural studies is often deliberate (and, as explained below, such studies often do resort to stylized ways of evoking or invoking temporal change). Perhaps the best thing to say in this connection is that I am remarking upon a hitherto unthematized feature of a genre that becomes clearly visible only from the standpoint of a developed analysis of practice.[7] Put more positively, while seeking in this section to establish an alliance with cultural studies of science, I also want to suggest that the latter inadvertently serves to efface a topic—the temporality of practice—that deserves more attention than it has so far received.

7. For example, Joseph Rouse has insisted to me that cultural studies of science *are* centrally concerned with change in time, but it is interesting to examine his important essay "What Are Cultural Studies of Scientific Knowledge?" (1992) in this connection. The closest he comes to defining "culture" is to say that "[t]he term 'culture' is deliberately chosen for both its heterogeneity (it can include 'material culture' as well as social practices [note the plural], linguistic traditions, or the constitution of identities, communities, and solidarities) and its connotation of structures or fields of meaning" (2). This is much like my own image of culture. Rouse then states that "I use the term [cultural studies of scientific knowledge] broadly to include various investigations of the practices through which scientific knowledge is articulated and maintained in specific cultural contexts, *and translated and extended into new contexts*" (2, emphasis added). It would seem, therefore, that Rouse is grouping what I call cultural studies under a single heading with what I call studies of practice (and certainly Rouse's own philosophical writings in this area display a developed interest in cultural transformations in time: see Rouse 1994). However, the body of his essay consists in developing six themes that he takes to characterize cultural studies of science, namely, "antiessentialism about science; a nonexplanatory engagement with scientific practice; an emphasis upon the materiality of scientific knowledge [*sic*]; an even greater emphasis upon the cultural openness of scientific practice; subversion of, rather than opposition to, scientific realism or conceptions of science as 'value-neutral'; and a commitment to epistemic and political criticism from within the culture of science" (7). These themes are quite congenial to my analysis, but none of them speaks directly to the real-time transformation of culture. Only the reference to "cultural openness" appears to mark a point of intersection with the mangle, but in fact does not. Rather than evoking the open-ended extension of culture in time, it refers to transverse connections linking science to the extrascientific world: "cultural studies of scientific knowledge display a constant traffic across the boundaries that allegedly divide scientific communities ... from the rest of the culture" (13; see 12–17 for more discussion).

With these remarks in mind, I now want to put some flesh on my notion of cultural studies of science by citing some examples and approaches and discussing how they relate to the mangle. Speaking crudely, one can distinguish three categories of work within the broad field of cultural studies of science.[8] One is a historicizing approach, which takes the timeless categories of traditional philosophy—"reason," "objectivity," "representation"—and situates them with respect to particular times and places. Hacking's work (1982, 1992b, 1992c) on "styles of reasoning" can serve as our example.[9] Traditionally in philosophy of science, "reason" has been conceptualized as an invariant property of scientists down the ages, but Hacking argues that what counts as scientific reason has been different at different periods. Hacking (1992b), for instance, puts dates on the origins of the statistical style of reasoning—1640–93—and traces out its subsequent mutations. The relationship between such approaches in cultural studies and the mangle should be clear: in section 6.6, I emphasized the historicist appreciation of scientific culture that follows from the mangle, and here cultural studies offer historical exemplification of the point in respect of the central, presumptively ahistorical concepts of traditional philosophy.[10] Hence my idea that cultural studies can be both orthogonal and complementary to the mangle.

A second approach to cultural studies of science is one that, in effect, follows Michel Foucault (1979) in its interest in the specific human disciplines and practices that characterize and help to constitute given scientific cultures. Having already explicitly discussed discipline in section 3.5 and later chapters, not much remains to be said on this topic, except to make it clear just how cultural studies in this area reinforce the

8. My distinctions are indeed crude; there is considerable overlap between the approaches discussed below. Further, my citations to each approach are intended as examples of specific pieces of work that I associate with it; they are not supposed to be exhaustive. I should also emphasize that I focus in what follows on the empirical (historical and ethnographic) findings of cultural studies, since these are what matter for my argument. This is not to say, however, that the authors cited approach cultural studies in a narrowly empiricist spirit. All of them, I think, are interested in central questions in philosophy, social theory, and historiography.

9. For other historicizing accounts of scientific reason, see Ginzburg 1980, 1989; and Hannaway 1975. Daston 1993, Daston and Galison 1992, and Porter 1992a apply the same historicizing tactic to great effect with respect to scientific "objectivity"; and on "representation," see Beyerchen 1989; Lynch 1985a, 1985b, 1991b, 1993a; and Lynch and Woolgar 1990.

10. This category of cultural studies thus returns us to the mangling of reason, objectivity, and so on from an angle different from that pursued in section 6.4.

mangle. This is simple enough. On the one hand, the concern with disciplines immediately lines up these studies alongside the mangle on the terrain of performativity. On the other hand, studies like Sibum's and the others cited in section 3.5 make it clear that the regularization of human agency is reciprocally bound up with captures and framings of nonhuman agency, with specific machines and instruments. Such studies, then, both reinforce the shift to the performative idiom and, at the same time, document just the kinds of interactive stabilizations of human and nonhuman performances (and, of course, representational systems, social relations, etc.) that the mangle leads one to expect. They constitute, in other words, snapshots of heterogeneous manglings captured in flux.

The third approach to cultural studies that I want to mention resembles the second but is more wide-ranging in its scope. I am uncertain as to its historical origins, but Latour's injunction to "follow scientists and engineers around" (1987) could serve as its organizing principle, at least if we read Latour as recommending that we should track down the movements of scientists not through time but in cultural space, that we should trace out the interconnections of heterogeneous cultural elements and strata that they weave, or that others weave around them; that we should explore the ways in which particular machines, disciplines, styles of reasoning, conceptual systems, bodies of knowledge, social actors of different scales, the inside and the outside of the laboratory, and so forth, have been aligned at particular times and in particular places.[11] This has proved a very stimulating line of inquiry over the past decade, and is therefore worth dwelling upon at some length. I can review Davis Baird's study of analytical chemistry as a straightforward but illuminating example.[12]

In his "Analytical Chemistry and the 'Big' Scientific Instrumentation Revolution" (1993; page citations below are to this article), Baird contrasts two regimes in the history of analytical chemistry, one dominant

11. To return to my earlier remarks, it is worth noting how the centrality of the "network" metaphor, as well as the appeal to semiotics (section 1.3), serves to encourage a cultural-studies articulation of actor-network theory and to discourage a systematic interest in temporality.

12. For more examples, see Biagioli 1990a, 1990b; Galison 1990, 1991, 1993; Haraway 1985; Hughes 1983; Lenoir 1992; Miller 1992; Miller and O'Leary 1994; Miller and Rose 1993; Rose 1990; Rose and Miller 1992; Schaffer 1988, 1992a, 1992b, 1993a, 1993b; Shapin 1988b; Shapin and Schaffer 1985; Traweek 1988, 1992; Wise 1988, 1993, 1995; and Wise and Smith 1989–90.

up to around 1920, the other securely installed from, say, 1950 onward. Baird shows that the old, pre–1920 regime was distinctively chemical, dependent upon a knowledge of chemical properties to separate out and identify, qualitatively and quantitatively, the constituents of unknown substances. Baird stresses that the equipment used in the old regime was simple and relatively cheap—calibrated glassware (pipettes, burettes), balances, and so on. And then he goes on to contrast this situation with that which obtained in the 1950s. The new regime in analytical chemistry was characterized by the omnipresence of physical techniques and their associated instruments—"electron microscopy, tracer techniques, infrared spectrophotometry, X-ray diffraction, mass spectrometry, chemical microscopy, polarography, etc." (277). The material stratum of analytical chemistry was thus largely transformed in moving between the two regimes. And along with that went other changes. At the level of material agency, the performativity of the new instruments was such that analyses could be carried out faster, on smaller samples, and with greater precision (268). At the level of human agency and discipline, new training schemes based on new textbooks were devised. Concurrently, a general deskilling was evident, as university-trained analysts increasingly became managers of relatively untrained technicians; and when scientists engaged in instrument development, they drew rather upon the characteristic skills and techniques of physicists and electrical engineers than upon those definitive of chemistry. At the level of institutions and social relations, new companies came into being, devoted to the development and production of new instruments; the marketing of their products became important; and conference circuits grew up to link these products into the worlds of the academy, government, and the post–World War II military-industrial complex (288).

Several points can be made about this example. First, Baird is not interested here in detailed transformations of culture in time. He offers us snapshots—transverse cuts through cultural webs, before and, in more detail, after World War II. His essay is thus a work of cultural studies as I have defined it. Second, as I said of cultural studies of discipline, this more all-embracing class also moves onto the same performative terrain as the mangle, focusing on machines, instruments, production, and so on.[13] Again such studies work, in effect, to document just

13. Baird actually glosses his study as being about "a new kind of scientific knowledge, scientific instrumentation" (1993, 268). I thank him for several discussions on the

the kinds of heterogeneous cultural alignments (and shifts therein) that the mangle would lead us to expect. But, third, I should note a complicating factor. There evidently is a certain sense of time in Baird's essay, which is, in fact, typical in this respect of work in cultural studies. Baird's before-and-after snapshots are taken at different times, and juxtapose two very different regimes in the history of analytical chemistry. I am inclined to make two remarks on this. On the one hand, such displays of difference and discontinuity serve to bring home the lack of necessity—in other words, the historicity—of any given cultural regime.[14] Here, again, cultural studies and the mangle hang together and reinforce one another. On the other hand, of course, it is inevitable that processes of mangling are entirely effaced by such displays of difference. There is simply no space in accounts like Baird's for the temporally emergent dialectics of resistance and accommodation that have been of central concern in my preceding chapters. This, in the end, is the sense in which cultural studies have an at most vestigial sense of time compared with analyses of practice, and in which the mangle promises to supply what is lacking.[15]

So the upshot of this discussion is that cultural studies of science and

question of whether it is useful to treat material objects as knowledge; we continue to differ on this topic.

14. One can return here to the notion of nonrepresentational incommensurability mentioned at the end of section 6.2. For example, the reference to "revolution" in Baird's title indicates his conclusion that the two regimes in analytical chemistry served to define disjoint forms of scientific life, but, as he states, the move from one to the other "did not involve changes in theory" (1993, 267). For some examples of an emphasis on contrast and difference in cultural studies beyond science studies, see Baudrillard 1988a; Eisenstein 1983; Greenblatt 1991; Said 1978; and Taussig 1987. Baudrillard 1988a is interesting in its exemplification of the idea that two cultures (contemporary American and European) can be incommensurable inasmuch as the axes along which one can be perspicuously characterized may not be perspicuous for the other. The departure of this sense of incommensurability from traditional discussions of how knowledge latches onto the world is total.

15. This vestigial sense of time in cultural studies is probably sufficient to resolve my argument with Rouse about time and cultural studies of science (note 7 above); I want to stress that there is *much more* to be learned about practice than can possibly be got from contemplation of discontinuities between cultures separated in time (or space, for that matter). It is also worth noting that in cultural studies of difference and discontinuity it is easy enough to slip back into traditional explanatory schemes. Cultural studies of science, on my definition, can only register correlations and alignments between multiple elements and strata, but, in default of tracking the details of cultural evolution, the temptation is to unreflectively assign explanatory priority to some subset of those elements. Hence, for instance, the prevalence of constraint talk in the cultural-studies literature.

the mangle reinforce and interactively stabilize one another. They elaborate a shared vision of what scientific culture is like: performative, historical, heterogeneous, multiple, and multiply interlinked. It is therefore reasonable to see cultural studies and the mangle as delineating two perpendicular axes of the new synthesis in science studies, the mangle staking out the time dimension of scientific practice while cultural studies engage in transverse cultural mappings. This is the principal point I wanted to establish in this section. It is important to remark, however, that, despite their orthogonality, cultural studies and studies of practice are not mutually exclusive vectors of inquiry. One can do both at once. One can, that is, imagine a kind of moving cultural studies that would combine an interest in cultural mapping with my present interest in cultural transformation in time. Of course, the studies reported in earlier chapters already do this to a certain extent. The analyses of practice in part 1 themselves entailed explorations of scientific culture, albeit limited to those cultural elements foregrounded in particular passages of practice. But here a micro/macro distinction has to be acknowledged. As noted at the beginning of this chapter, even the study of N/C at GE has a fine-grained focus compared with cultural studies of science as just exemplified. The latter typically aim at considerably broader characterizations of culture than attempted anywhere in part 1. It thus remains an open question how one might integrate synchronic and diachronic concerns at the macrocultural level, and in sections 7.3 and 7.4 I will address this question by talking about the kind of macrohistoriography of science that the mangle invites. First, though, one further concern with cultural studies also informs subsequent discussion.

"Cultural studies" is not a neutral term. It already functions as the name of an exciting and often controversial field of nontraditional approaches in the social sciences and humanities in general. And I should admit that I have been talking about "cultural studies of science" as an attempt to align them, and hence the mangle, with what I take to be a much wider field of antidisciplinary synthesis.[16] I wanted to suggest that cultural studies of science can be seen as part of cultural studies more generally, and that perhaps my analysis of practice conceptualizes the temporality of cultural transformation in a way that is underdeveloped

16. On antidisciplinarity, see the opening quotation to this section from Stefan Collini, which continues, "One ingredient that is common to [cultural studies] is discontent with what are perceived to be the limitations and obstructiveness of the institutionally well-established academic disciplines" (1994, 3).

in both. But now I have to face the question, are cultural studies of science, as I have defined them, really part of cultural studies at large? Especially, since my concern here is with time, are the latter well characterized by a focus on mapping cultural webs rather than by the analysis of real-time practice?

These are not easy questions to answer, not least because, if my characterization of cultural studies of science is contentious, any characterization of cultural studies *tout court* is even more so. One can look to the thirty-nine essays contributed to a massive volume on cultural studies for a cross-section of the field, but, as the editors themselves remark, "it is probably impossible to agree on any essential definition or unique narrative of cultural studies" (Grossberg, Nelson, and Treichler 1992, 3). It is relevant to note that the editorial introduction to that volume repeatedly instantiates an implicit definition of culture congruent with that given above, being concerned with transverse connections across multiple and heterogeneous fields and evincing little interest in processes of transformation in time.[17] But rather than pursue that line of thought, I want to close this section with one more example. At the moment, Donna Haraway is probably the key figure in the articulation of science studies and cultural studies, and I want to make some points about her "Manifesto for Cyborgs" (1985)—one of her best-known essays, much discussed within and beyond science studies—taking it as an exemplar of cultural studies "proper."[18]

The first point to emphasize about Haraway's "Manifesto" is that its

17. For another crosscut through cultural studies that points to a similar conclusion, one can scan the ten book reviews and associated advertising in a long section devoted to cultural studies in the *Times Literary Supplement* (1994). In the lead essay, Collini first notes that "the term [cultural studies] is used so variously" that an attempt to define it by its "subject-matter and methods ... seems to me doomed to failure." But later he gives, in effect, his own definition: "we are not dealing with 'culture' in the singular at all, but with that plurality of symbolic systems and practices that enable different groups to make various kinds of sense of their lives. This usage clearly owes more to anthropology than to Arnold" (3–4). In the same vein, one can note that Jameson concludes his influential essay "Postmodernism, or The Cultural Logic of Late Capitalism" with the argument that "a model of political culture appropriate to our own situation will necessarily have to raise spatial issues as its fundamental organizing concern ... The political form of postmodernism, if there ever is any, will have as its vocation the invention and projection of a global cognitive mapping, on a social as well as a spatial scale" (1984, 51, 54). On the politics of cultural studies, see below.

18. Though even this is contentious, I would say that Haraway 1992 is the only contribution from a "science studies" author in Grossberg, Nelson, and Treichler 1992; see also Haraway 1991a.

central sections (161–73) are structurally isomorphous with Baird's essay just discussed, though much more wide-ranging, scientifically, technologically, and socially. Like Baird, Haraway contrasts two cultural formations, which, as it happens, can again be labelled as pre– and post–World War II, or, in Haraway's case, modern and postmodern.[19] In the "Manifesto," Haraway is principally concerned with the second of these, and much of her text consists in "mapping" (150) the novel forms and alignments of multiple and heterogeneous elements and strata that characterize it: microelectronics, communications technologies, biotechnology, "modern states, multinational corporations, military power, welfare state apparatuses, satellite systems, political processes, fabrication of our imaginations, labour-control systems, medical constructions of our bodies, commercial pornography, the international division of labour, . . . religious evangelism depend[ent] upon electronics" (167), the "homework economy" and the "feminisation of labour" (166–69), "women in the integrated circuit" and the new configurations of "home," "market," "paid work place," "state," "school," and "church" (170–73), and so on. This is not the place to enter into the specifics of Haraway's analysis of postmodern culture; what I want to emphasize is that, again like Baird's, it is carried through in a performative idiom. The whole point of Haraway's "cyborg" metaphor, which runs through

19. Haraway opens the sections in question with the assertion that "we are living through a movement from an organic, industrial society to a polymorphous, information system—from all work to all play, a deadly game." Like Rouse (n. 7 above), she thus appears to be concerned with real-time practice, but she continues, "the *dichotomies* may be expressed in the following chart of transitions from the comfortable old hierarchical dominations to the scary new networks I have called the informatics of domination" (emphasis added). The chart takes the form of two columns in which the tension between corresponding terms evokes the dichotomous transitions just mentioned. It covers almost a page, and its first few entries are "representation," "bourgeois novel, realism," "organism," "depth, integrity," "heat," in the left-hand column; and "simulation," "science fiction, postmodernism," "biotic component," "surface, boundary," "noise," in the right-hand column (161–62). The columns are not labelled in the "Manifesto," but a paragraph on the following page (163) begins by associating the break between them with the Second World War and ends by associating the second column with postmodernity. The importance of this dichotomous chart to Haraway's thought is evident in that it appears (with minor transformations over the period 1979–89) in no less than three of the essays collected in Haraway 1991b (44, 161–62, 209–10). In its earliest appearance, the columns are labelled pre– and post–Second World War (44). In the "Manifesto," Haraway immediately uses this chart to stress the historicity of culture: "the objects on the right-hand side cannot be coded as 'natural,' a realization that subverts naturalistic coding for the left-hand side as well" (162).

and organizes the "Manifesto," is (in my terms) to thematize the reciprocal interdefinition of human and nonhuman agency, of machines, instruments, technologies and their performances, human disciplines and practices, the scale and relation of social agents, and so forth. For Haraway, culture (at least, postmodern culture) is a cyborg formation, itself "a cybernetic organism, a hybrid of machine and organism . . . simultaneously animal and machine," living in a world that is "ambiguously natural and crafted" (149).[20]

So the "Manifesto" is a work of cultural studies of science as I defined and discussed that field above, and if one is prepared as well to regard it as an exemplary work of cultural studies *tout court,* then one can see both how cultural studies of science can shade into a wider antidisciplinary synthesis under this heading, and, by implication, how the mangle fits in and promises to contribute to the overall picture in the delineation of a temporal axis for analysis. This is, I think, an important way in which the long-standing isolation of science studies, mentioned at the beginning of section 7.1, can be overcome.[21] But one point still remains to be clarified.

As its name suggests, Haraway's "Manifesto" is both a description of a situation and a call to arms. Its opening and closing sections (149–61, 173–81) are devoted to issues of political theory and practice. Haraway would like to transform contemporary culture in a "socialist-feminist" direction (as the subtitle of her essay states), and her cyborg metaphor helps her to think this through along nonessentialist—posthumanist—lines. Such an explicit political commitment is, I must acknowledge, often taken as definitive of cultural studies (see, once more, the introduction to Grossberg, Nelson, and Treichler 1992; and Collini 1994, 4). And if one accepts this definition, then, on the face of it, neither the mangle nor much of what I have been calling cultural studies of science

20. One complication enters here. Haraway uses the figure of the cyborg to signal the intensity and self-awareness of human/nonhuman couplings and fusion that she takes to characterize the postwar era, rather than culture in general. Elsewhere, however, she is more symmetrical in her attention to pre- and postwar cultural formations, and identifies performative alignments of human and nonhuman agency in both (Haraway 1979). Echoing Latour (1993b), one could paraphrase her analyses as showing that we have always been cyborgs but never noticed it before.

21. And perhaps cultural studies might be able to learn something from the process: much writing in cultural studies, unlike Haraway's, continues to operate in a representational idiom (see, for example, the contributions to Penley and Ross 1991 for the translation of mainstream cultural-studies concerns into the realm of technology).

has the proper credentials to be a full part of the wider synthesis at issue. Neither Baird, say, nor I hitch our analyses to specific political projects.[22] But there is, I believe, more to think about here, and the following observations might help to elucidate the sensibilities behind sections 7.3 and 7.4.[23]

I do not think that analysis and politics are necessarily as closely linked as, for example, Haraway's "Manifesto" suggests. While Haraway sees her cyborg imagery and her politics as tied together, this book, after all, has arrived at notions of the coupling of human and nonhuman agency quite congruent with Haraway's, but via a consideration of the temporality of practice rather than through any commitment to nonessentialist socialist-feminism (though, as it happens, I would be happy enough to support Haraway's agendas).[24] Nevertheless, cultural studies of science and the mangle can be made to intersect with political debate. Thus, along their different axes, both serve to destabilize the nonemergent and humanist premises of traditional political thought. The posthumanist perspective that emerges from my analysis of practice, for instance, tends to undermine any faith in a distinctively humanist politics; it reinforces, to put it the other way around, political programs that explicitly aim at symmetrically interlinked transformations in the human

22. Latour (1993b) does conclude with political discussion. He argues that we need a "Parliament of Things," but "we do not have to create this Parliament out of whole cloth, by calling for yet another revolution. We simply have to ratify what we have always done, provided that we reconsider our past, provided that we understand retrospectively to what extent we have never been modern, and provided we rejoin the two halves of the symbol broken by Hobbes and Boyle" (144). Latour is never, however, very specific on what it is that "we have always done," and there are plenty of reasons for hesitation before "ratifying" this blank check. This is where cultural studies of science can play an important role in political thought, as discussed below.

23. I thank Barbara Herrnstein Smith for discussions on what follows.

24. Cultural analysis and politics are treated as one thing in the opening sentences of the "Manifesto," which Haraway sets out as "an effort to build an ironic political myth faithful to feminism, socialism, and materialism... Irony is about humor and serious play. It is also a rhetorical strategy and a political method, one I would like to see more honored within socialist-feminism. At the center of my ironic faith, my blasphemy, is the image of the cyborg" (149). I have not singled out feminist scholarship for explicit discussion in this book because my examples promise to contribute little to it and, conversely, because I have found little on practice, as I define it, within the feminist canon. One can consult Rouse 1992 for access to feminist contributions to cultural studies of science. I thank John Law for pointing out to me that in the "Manifesto" the figure of the cyborg is also used to evoke a second theme (besides posthumanism) developed by feminist scholars, that of "fractured identities" (155–61)—a multiplicity and decentering of selves and social formations. I do not address this topic for the reasons just given, but see also Graham 1994; Mol 1991; Singleton and Michael 1993; Star 1991a; and Strathern 1991.

and social, scientific, technological, and material spaces we inhabit. Likewise, the temporal emergence of the mangle suggests that whatever political agendas we construct should be situated ones, addressed to specific cultural formations and aspirations rather than founded on non-emergent, master-narrative construals thereof (Smith 1988, chap. 7; 1992). Assessing where we are now is thus a part of political practice, even if it cannot in itself issue in well-defined programs of action, and this is surely one of the impulses behind cultural studies of science.[25] In these respects, then, it seems to me that cultural studies of science and the mangle can enter into the field of political concerns even when not explicitly connected to definite agendas, and in this sense they do stand as contributions to the wider antidisciplinary synthesis of which cultural studies is a major nexus.

Of course, it must be conceded that the studies of part 1 of this book, now thought of in their microcultural aspect mentioned above, are not very perspicuous in characterizing the wider cultures in which they are situated. So, as promised already, in sections 7.3 and 7.4 I turn to the question of how my analysis of practice might be brought to bear on macrocultural formations and their transformation—as a topic of interest in its own right and as a contribution to political thought.

7.3 PERFORMATIVITY AND HISTORIOGRAPHY: THE BIG PICTURE

> To sloganize: the socio-logic of science as action deconstructs rational and ideological reconstruction. The possibility of producing a big picture for 2001 . . . nevertheless rests best upon gearing this shift to produce a narrative whose specific temporality is the time of science's worldly power.
>
> J. R. R. Christie, "Aurora, Nemesis, and Clio"

So far I have treated the history of science as a source of exemplifications of the mangle, taking them where I could find them. There are many reasons, however, for treating history with more respect, for an interest in historical specificity. As just discussed, one might want to interrogate

25. Baird's designation of the transition between the old and new regimes of analytical chemistry as a "big revolution," for instance, suggests both that this particular cultural transit is typical of those made by many sciences en route to their present forms and that recognition of this is important in political thought about the contemporary configuration of STS (Baird 1993, 288–89; the idea of a "big revolution" originates with Hacking 1987).

history as a way of getting to grips with the present, a way of analyzing the constitution of the world we live in and of arriving at a critical appreciation of its singularity. I want now to consider how the mangle might play into such concerns. There are many levels at which one could approach this topic, but here, as announced, I make the move from the micro to the macro, and ask what big thematic organization my analysis of practice suggests for the historiography of science. To echo the title of a recent collection of essays, *The Big Picture* (Secord 1993), which grew out of a conference by the same name, I am interested in exploring what kind of a big picture of the history of science (and technology and society) the mangle might sustain and be sustained by. One might think that as a general account my analysis has nothing in particular to tell us, but in this section, I want to show that the move from the representational to the performative idiom can generate some interesting lines of thought. In section 7.4, I will return to the mangle more specifically.

One can reason like this: The representational idiom fosters a certain line of historiographic attack on science. It encourages us to see periods of rapid representational change as key transits on the road to the present, and thus as key foci for historical research. By compiling a history of the Scientific Revolution, the Darwinian Revolution, the revolutions of relativity and quantum mechanics—all understood as breaks primarily in theory—we can construct a perceptive genealogy of our present world of representation. And this is how the history of science has traditionally been organized. In the performative idiom, however, we can follow the same line of thought to a different destination. The performative idiom encourages us to carry out a genealogy organized around striking transformations in the realm of human and material performances, which for me, to put it simply, would foreground not the Scientific Revolution of the seventeenth century but the Industrial Revolution of the late eighteenth and early nineteenth centuries as a key moment in the history of the West—the time when machinic performativity, as enshrined in the factory and in the distinctively disciplined human performances associated with the factory, started to become definitive of society itself.[26] The performative idiom, then, invites a *performative*

26. In the *Big Picture* volume, Cunningham and Williams suggest that we should shift the dates of the scientific revolution itself toward the present, arguing that "the period 1760–1848 is a much more convincing place to locate the invention of science" (1993, 410), and Secord's introduction observes that "several of the following essays point to the central importance of the decades around 1800 for the sciences, both in Europe and in terms of their export to other parts of the world" (1993, 388). "Unfortunately," he contin-

historiography, one might say, that would be centered in the industrial era on technology, the factory, and production (and consumption and also destruction, as mentioned below).[27] I can straightaway observe that such a performative historiography would necessarily abandon the traditional assumption that the history of science can in general be conceptualized as a self-enclosed field, substituting for it an interest in STS as a field of multiply crosscutting relations. This, I think, is another important respect in which traditional disciplinary boundaries around inquiry start to break down within the new synthesis. But now I need to make more precise what the move to a performative historiography might entail.[28]

First, I should acknowledge the close relation between performative and Marxist historiography. Marxism, after all, fixes upon production as the key moment in human activity, and upon the factory as the key site in making the modern world. And Marxist historians of science like J. D. Bernal (1953a; 1953b; 1969, vol. 2) were among the first to explore the intertwining of modern science with industry and capital. But the kind of performative historiography that I have in mind differs from its Marxist relatives inasmuch as the latter are typically constructed around traditional humanist and temporally nonemergent master narratives. Bernal himself, for example, took for granted a very traditional understanding of science as a method of getting at the unique objective

ues, "the significance of this period is scarcely evident to anyone who approaches the literature for the first time. The fact has to be teased out from a dozen histories of chemistry, geology, physiology [and so forth]."

27. Christie (1993, 404) likewise recommends "a shift from an epistemological to a performative conception" in the historiography of science, and his entire essay, which I encountered only after writing this section, is very insightful on the promise, theoretical implications, and difficulties of this shift.

28. I should make clear that, while I concentrate on the industrial era in what follows, I do not want to suggest that a performative historiography of science can begin only with the Industrial Revolution. By definition, an organizing focus on industrial machinery and the factory would be impossible before that, but it would still be perspicuous to think through the history of science and society in terms of practical struggles in fields of human and nonhuman agency (in respect of housing, farming, fishing, shipping, warfare, and so on). It would require another book to develop this thought, but Bernal (1969) offers a preliminary map of the terrain. It also appears that one can get at much of the history of science in the preindustrial era via a focus on military machines and warfare. Voss (1992), for example, analyses the emergence of physics from the sixteenth-century intersection of practical mathematics and bombardiering. As Mahoney puts it, "It is a matter of emphases rather than alternatives, but one will understand Galileo's new sciences best by looking not at Plato's Academy, or even at the Accademia dei Lincei, but at the Arsenal of Venice" (forthcoming, p. 5 of draft).

truth of nature. On his view, social, political, and economic contexts could influence the rate and direction of scientific progress, but not the contents of scientific knowledge. On the social side, traditional Marxisms have envisaged a teleological progression of social forms, destined to arrive at socialism itself (Hessen 1932 is the classic statement). Evidently, the mangle undercuts such accounts of science and society. The relativism and historicism of the mangle (sections 6.5 and 6.6) undermine the idea that science gets at unique truths of nature independently of context and contingency or that the vectors of social transformation are somehow given in advance of practice, and I argued in detail against a traditional Marxist theorizing of the social in section 5.3.

So the performative historiography that the mangle suggests departs from traditional Marxism in stripping the latter of its master-narrative structure. Like Marxism, it suggests an organization of historical study around sites of machinic performativity. Unlike Marxism, though, it suggests that we see the factory, say, not as the focal point for the playing out of preordained historical scripts but as a *double surface of emergence,* for science on the one side and, as in chapter 5, for society on the other (and, of course, as a locus for continual reencounters of the two). It suggests, in other words, that as historians our business might be to explore open-ended transformations of science and society in terms of the temporally emergent making and breaking of cultural alignments and associations with the worlds of production and consumption, transformations understood as having no determinate destination in advance of practice. Leaving aside the transformation of the social for the moment, the idea would be, for instance, to explore how existing sciences were decisively inflected by their encounters with industrialization in the nineteenth century and into the twentieth, and with how new sciences came into existence in industrial settings. Classic instances here (all mentioned in Bernal 1953a) include the construction of the modern science of thermodynamics in the field of industrial concerns surrounding the steam engine, of electromagnetism around the telegraph, and of microbiology around the industrialization of food and drink production.[29]

29. Once one begins to appreciate the multiplicity of links between the modern sciences and industry, it becomes clear that "pure science" is far from science's natural state (contrary to the standard assumption in representationalist history of science). Pure science, inasmuch as it exists, has instead to be seen as a specific historic achievement. The founding of CERN, the European high-energy physics laboratory, is an amusing and almost paradoxical instance of how science has occasionally been purified. At its inception in 1954, CERN's research agenda was explicitly defined as decoupled from economic and

It would be exciting and perspicuous to carry through such a performative historiography of science. It would, I think, contribute importantly to a critical grasp of the contemporary world. And, of course, there is a growing historical literature organized along the lines just suggested. It remains the case, however, that this body of literature is tiny in comparison with the established representationalist historiography of science (which also continues to grow). It is also the case that, beyond the traditional Marxist oeuvre, little attention has yet been paid to synthesis in performative historiography: I cannot think of a single book that offers an overview of the emergence of the contemporary sciences in relation to economic production without leaning on the usual master narratives of truth, class interests, or whatever.[30] The point that I want to emphasize, therefore, is that the mangle supports and encourages the continued growth of a performative historiography of science by showing how it is possible to escape from and transcend the representational idiom and to conceive of both science and society as genuinely emergent in a field of agency and performativity. Performative historiography is (or would be) the substantive counterpart to the analytic shift to the performative idiom that I argued for in chapter 1; it is a rebalancing of our historical understanding that reinforces and is reinforced by the mangle as a general account of practice. To put the point as forcefully as I can: from the perspective of the mangle, any attempt to understand the overall history of science as that of a self-contained field of knowledge and representation seems almost obscenely skewed, an apparently

military concerns as a condition for international political support, not so much because the nations of Europe wanted to support pure science, but because in the period of postwar reconstruction they wanted to collaborate in some international scientific venture without conceding economic and military advantages to one another (Hermann et al. 1987; Pestre and Krige 1992). The cancellation of the US superconducting supercollider in 1993 is a token of how fragile an achievement purity is, while the coincidence of this cancellation and the end of the Cold War suggests that purity is at most relative.

30. For some examples of performative historiography in the cultural-studies mode, see the works on the nineteenth and twentieth centuries cited in notes 9 and 12 above. Undoubtedly a considerable literature could be recruited to the cause of performative historiography from existing bodies of work in the history of science, technology, business, warfare, and so on, but how the pieces might connect together, and what would remain after stripping them of taken-for-granted traditional explanatory structures, remains to be thought through. It is, of course, the case that synthetic history of science as a genre has fallen into desuetude. As Christie observes, "[T]here exist no large-scale syntheses of history of science which integrate the empirical and theoretical developments of historiography since 1962" (1993, 400); he then goes on to discuss the practical difficulties entailed in the construction of a performative big picture—one has to know about an awful lot of heterogeneous topics (seen from the perspective of the traditional disciplines).

deliberate denial of the performative contours, human and nonhuman, of our contemporary condition and its historical antecedents.[31]

7.4 MACROMANGLING

Suppose we considered the war itself as a *laboratory?*

Thomas Pynchon, *Gravity's Rainbow*

What would a performative history of STS in the large look like? How would one go about conceptualizing it, organizing it, grasping it in detail? One answer, legitimated by the studies in part 1, would be that it would take the form of an enormous assemblage of interacting material-conceptual-social-etc. manglings at the microlevel of individual and small-group practice. And, actually, I think this answer would be right. I cannot imagine that history happens otherwise. But it is also a useless answer, inasmuch as one could never grasp, at that level of detail, the history of industrialization, say, and the associated transformations of science and society. It is at this point that the yearning for the traditional humanist and nonemergent master narratives of history returns, the desire for some substantive organizing principle, however flawed. Fortunately, however, there is another answer to my question. It seems to me that the mangle is, as the physicists put it, *scale invariant:* at whatever level of magnification one interrogates history, one finds mangling. At the microlevel (chapters 2 to 4), one finds micromangling; at the macrolevel, one finds macromangling. My analysis of practice, in other words, carries over directly to the practice of macroactors.[32] That, at any rate, is the claim I want to exemplify in this section.

31. And here, alas, I stand condemned by my own words, at least in one respect. Chapters 2 to 4 (though not chapter 5) center on examples taken from the history of pure science, and the machines and instruments that are the focus of chapters 2 and 3 (though, again, not of chapter 5) are peculiar ones from the perspective just arrived at, directly aligned with the world of scientific representation and knowledge, not with the world of production. It is clear that when I began the project that has culminated in this book I, too, was under the spell of representation. In my own defense, I can only remark that the shift to the performative idiom has, for me, grown out of the studies of part 1. The need for the shift is a conclusion of my researches in the very heartland of representationalist historiography, not a presumption of them.

32. Callon and Latour (1981) argue to similar effect in respect of actor-network theory (and see also Knorr-Cetina 1988). Ideas of scale invariance seem to be becoming popular throughout science studies, often finding expression in terms of chaos theory and fractal imagery. Thus Jed Buchwald once described the contents of Smith and Wise 1989 to me as "fractal history," Piet Hut has talked about a "fractal Kuhn," and a journal recently

As my example, I could take E. P. Thompson's wonderful study *The Making of the English Working Class* (1963; see also Linebaugh 1992). At the very inception of the Industrial Revolution, Thompson demonstrates the open-ended transformation of English social structure around the factory and industrial capital. He shows how the working class constituted itself—delineating its own boundaries, constructing its own characteristic institutions, articulating its own interests, making itself a macroactor—in a temporally emergent dialectic of resistance and accommodation with machines, architectures, factory owners, the state, the church. Instead, however, I want to focus on the more recent example of the historic encounter of science and the military in World War II—in part because it is an instance of the increasingly complex intertwinings of science, technology, and the social that have taken place since the Industrial Revolution, and in part because a shift of focus from the factory to war can serve to register the central place that destructive as well as productive performativity must occupy in a performative historiography of the modern world.[33] In what follows, I concentrate on wartime developments in the US, and my text for this last exemplification of the mangle is Daniel Kevles's study *The Physicists* (1987, chaps. 19, 20; see also Pickering 1993a). My intention is especially to thematize open-ended transformations of the social (in contrast to the traditional assumption of stable actors and their properties) since this, I think, is a key area for concern in moving from the micro- to the macrolevel.

As a baseline, we can note that the prewar relationship between science and the military in the US was tenuous; the two institutions were more or less decoupled. The armed services did have their own technical bureaus, but their effectiveness was circumscribed by "small budgets, lack of interservice cooperation, and limited contacts with civilian science" (Kevles 1987, 290; chapter and page citations below are to this book). In the eighteen months before the US entered the war in December 1941, however, this situation had already begun to change. The impetus for the transformation came from civilian scientists who felt that they had more to contribute to the war effort than the military recog-

arrived in my mailbox has an article entitled "Philosophical Fractals, or History as Metaphilosophy" (Murphy 1993).

33. Note that the works by Baird and Haraway discussed in section 7.2 locate their dichotomous cultural transitions at World War II. Kevles and Geison (1995) stress the constitutive role of both world wars in the development of the experimental life sciences in the twentieth century.

nized, and its institutional vehicles were first the National Defense Research Committee (NDRC) and then the more powerful Office of Scientific Research and Development (OSRD). Modelled on the interwar National Advisory Committee for Aeronautics, the NDRC and OSRD sought, by letting contracts, to reorient the practice of civilian scientists—especially physicists—toward the production of hardware that would be useful to the armed services. In my terms, the aim was *to tune the practice of science as a social macroactor* toward particular captures of material agency. In the event, the most important machines and devices to emerge from this tuning were radar sets and atom bombs, and here I will concentrate on the former (chap. 20).

The central site for the development of radar in the US was the new Radiation Laboratory, or Rad Lab, established by the NDRC at MIT, and two points about the Rad Lab are worth noting. First, its mission was, as just indicated, the tuning for military ends of particular material devices, starting with the magnetron power source. Second, this material tuning was accompanied by a further tuning of science as a social institution, as a macroactor. The Rad Lab was a key site of propagation for the big-science work style—object-oriented and characterized by large, heavily funded, hierarchically organized, interdisciplinary teams of scientists, engineers, and technicians (already mentioned in chapter 2).[34] Part of the business of aligning scientific and military enterprise was, then, a transformation in the discipline of doing science itself, *in the very identity of science* as a macroactor. We can also note that the flourishing of neither the radar project nor the Rad Lab as an institution was guaranteed in advance. Early attempts to develop the AI–10 aircraft-detection radar were rejected by the US armed services. The use of radar, however, had contingently proved effective in actions against German U-boats around the British Isles, and when the Rad Lab switched its efforts to antisubmarine radar in the air-to-surface-vessel (ASV) project, the navy actually ordered sets for experimental and operational use (305). This completed the circuit linking the machines developed by civilian scientists into military use, and guaranteed the future of the Rad

34. "By the end of 1942 the Rad Lab . . . budget had reached $1,150,000 monthly, and its staff had multiplied to almost two thousand people. By 1945 it would contain almost four thousand, about one quarter of them academics, almost five hundred of them physicists" (307). E.O. Lawrence's prewar Radiation Laboratory at the University of California in Berkeley was an influential model in the World War II proliferation of big science. See Heilbron and Seidel 1990 on Lawrence's lab, and Galison and Hevly 1992 on the history of big science more generally.

Lab. Here, then, we have a classic example of macromangling in a heterogeneous culture: in the open-ended space of cultural extension, a particular development of science as a macroactor (big science as instantiated at the Rad Lab) and a particular extension of prior machinic culture (the ASV radar set) were interactively stabilized against one another in a project oriented to the capture of material agency for military purposes.

With ASV and other successes, in radar and elsewhere, the scientists began to press—against military resistance—for a more intimate engagement in military affairs. Their argument was that scientists should be allowed to make their own assessment of military needs, and that they should be allowed to involve themselves in the conduct of military operations. Both were necessary if the civilian development and military use of new hardware were to be optimized (tuned to one another). At a high level of authority, the first objective was initially met in the establishment of the Joint Committee on New Weapons and Equipment (JNW). Composed of a rear admiral, a brigadier general, and, as chair, a civilian scientist (Vannevar Bush, the head of the OSRD), the JNW was both a device for the fusion of civilian science and the military and a kind of double panopticon, intended to provide visibility for each partner into the workings of the other.[35] In embryo at least, then, the nature of both civilian science and the US armed services was transformed by the establishment of the JNW: each *opened* itself to an unprecedented extent to the other as a surface of emergence.

As far as the military use of new science-based technologies like radar was concerned, a key transformation was centered on the development of operations research (OR; for more on the history of OR, see Morse 1977; and Fortun and Schweber 1993). As its name suggests, OR was a conceptual approach developed by physicists and mathematicians for the optimization of military operations. It was, one might say, a strategy for tuning the performativity of the military machine. Early applications of OR focused on finding such things as "the optimal bomber formation

35. Earlier such fusions were evident in the copresence of military and civilian members in the NDRC and, at a practical level, in the fact that "[a]fter Pearl Harbor and the success of the ASV . . . scores of army and navy officers established residence at MIT to keep their respective services up to date on the military possibilities of microwave systems and to inform the physicists of current military needs" (306). "Panopticon" is, of course, a gesture toward Foucault's discussion (1979) of one-way surveillance as a technique of power. The interesting thing about the JNW is that it worked both ways, hence double panopticism.

for minimising losses from enemy flak and fighters, or the optimal search and attack pattern for going after a submarine hidden in the vast expanse of the sea" (311). Several features of the development of OR in World War II are worth noting. First, it often implied changes in military tactics—a shift toward aggressive air searches for submarines, for example, and away from an absolute reliance on the primarily defensive use of naval convoys. It implied, then, a transformation of the nature of military enterprise itself (and was for a time resisted as such; 313). Second, the effective utilization of OR techniques depended on a thoroughgoing integration of civilian OR scientists into the military body. For example, in July 1943, the navy was persuaded "to incorporate into the Tenth Fleet the group of operations research scientists that had originated in Boston" (314).[36] There, like other subsequently incorporated OR groups, the scientists and mathematicians had access to operational data, and were involved in the planning and evaluation of specific operations, reporting directly to commanding military officers.

Beyond opening itself to science, then, and beyond a limited fusion on the neutral ground of newly established committees and the like, here the military *enfolded* civilian science, reconfiguring its internal structure to embrace OR practitioners and reconfiguring its operations around their findings and recommendations. Again, this enfolding amounted to a significant transformation of the military as a social macroactor. And again, this transformation proved performatively effective—"[i]n April 1943 convoy losses dropped sharply" (315)—leading to an intensification of the enfolding of science by the military.

> [I]n January 1943 there had been only one operations analysis section in the entire U.S. Army Air Force; by January 1945, in USAAF commands around the world, there were seventeen such groups, employing 32 mathematicians, 21 radio and radar engineers, 14 terminal ballisticians, 11 physicists, and some 100 other analytic experts . . . By V-J Day, from Africa to Southeast Asia and on to the Aleutians, civilian scientists were in vogue as strategic and operational advisers to a degree without precedent in the annals of American military history. (320)

In these developments, we thus have another exemplification of macromangling, this time of the material, the social, and the conceptual.

36. As Philip Morse, the leader of the Naval Antisubmarine Warfare Operations Research Group (ASWORG), put it, by 1942, "we were into the Navy more deeply than anyone had thought possible, and I felt we could penetrate more deeply still" (1977, 187).

Particular transformations of disciplined military agency (the shift in tactics from defense to offense in antisubmarine warfare) and particular transformations in the social relations of science and the military (the opening up of preexisting boundaries between them, the enfolding and incorporation of the former by the latter) hung together with and were interactively stabilized alongside the material performativity of radar and the newly constructed conceptual technology of OR. One last point. The transformations wrought on the military body by the enfolded scientists were not limited to the conduct of specific operations. To give just one example, specialists reviewing radar-guided bombing operations noted that some bombing groups consistently performed better than others, an observation that they interpreted as indicating that "B–17 bomber crews had to be better trained and conditioned in a variety of ways" (319). Part of the tuning of the military that went with the enfolding of science was, then, the construction of new military disciplines designed to support the daily use of new science-based technologies.

The intersection of science and the military in World War II can thus be understood as a macromangling that encompassed both an inner transformation of these two macroactors and an outer transformation in their relations to one another. The way of doing science changed from small to big science; the military shifted its tactics and basic disciplines; both institutions were topologically transformed in a reciprocal transformation of shape marked by the opening of boundaries, fusions in newly created institutional spaces, the creation of a new and reciprocal optics between them, and the enfolding of one macroactor by the other; and all of these transformations were interactively stabilized in relation to transformations in machinic culture (symbolized here by developments in radar technology) and in conceptual apparatuses (OR). What we have in this example, therefore, is a story of cultural mangling analytically isomorphous with the material-conceptual-social manglings we examined in part 1, but writ large, in terms of social macroactors and classes of material and conceptual technologies.[37]

37. To return to a remark made much earlier (chap. 1, n. 18), I do not think that the "network" metaphor of actor-network theory is adequate to grasp the topological transformations just summarized. "Enfolding," especially, cannot be adequately conceptualized as a linear connection of nodes. One might also comment here on the problems that episodes like that under discussion raise for "contextualism," the idea that one can overcome the traditional internal-external distinction in the historiography of science by appealing to the wider context in which scientific developments are situated. In this in-

So much for my overview of the shifting contours and relations of science and the military in World War II. It is, of course, no more than a thumbnail sketch. One could do more toward establishing the open-endedness of the cultural extensions at issue by looking at material, conceptual, and social vectors that did not get stabilized, and the overall picture could be refined, expanded, and continued toward the present in all sorts of ways. I want to stop here, though, and review where this and section 7.3 have led us.

My interest has been in how the mangle might engage with concerns for understanding STS in the large. In section 7.3, I talked about a shift to a performative historiography that my analysis of practice might help to reinforce, a historiography organized around technology and the factory as a center of production and, as I have sought to make clear above, around military technology and military enterprise as a center of destruction. In this section, I have been inquiring into how we might conceptualize such a historiography at the macrolevel, and through my World War II example, I have sought to demonstrate the scale invariance of my analysis of practice. I have sought, that is, to show that the mangle can illuminate cultural extension at the macrolevel as well as at the finer levels of detail at issue in part 1. My intention has been to show that, contra traditional social-theoretic accounts in terms of stable actors possessing stable properties, the nature of science and the military as macroactors was emergently transformed in their World War II encounter, and that this transformation was itself stabilized in a posthuman fashion alongside new material and conceptual technologies like radar sets and OR. Macromangling, then, is my suggestion for how we might think the macrohistory of science, of technology, of society, once we relinquish the comforting but untenable analytic blueprints of traditional philosophy of science and social theory—for how we might construct a historical big picture without appealing to (humanist and nonemergent) grand narratives.

I close with two remarks. First, though I have exemplified the scale invariance of my analysis with a single example of macromangling, I think it is clear that this idea can be taken much further. I have already

stance, the coupling between science and the military was so strong, and the reciprocal transformations thus wrought on each were so far reaching, that it makes little sense to see either as offering a contextual explanation for the development of the other. For more on this, and on the gestalt switches precipitated by contextualist historiography of strongly coupled developments, see Pickering 1992a.

noted, for example, that E. P. Thompson's work can be read as an account of the early mangling of the macrostructure of industrial society; and, to remain in the domain of the factory, there are ample grounds for thinking that the reorganization of science, technology, and production around industrial and governmental research laboratories, which began in the late nineteenth century and continues to the present, has much the same character as the scientific/military developments just discussed. The novel space of the industrial research laboratory is manifestly a site at which the practice of science has been systematically tuned to industrial ends, and at which industry has systematically enfolded science—reconfiguring its inner constitution to make a space for, and to optimize its capture of, the agency of science, scientists, and their machines.[38] In the present, one thinks of reciprocal and contested tunings of social forms and computer networks. The mangle can, then, go a long way in helping us to conceptualize large-scale historical developments.

My second remark returns to my earlier ones about historical and cultural specificity. Traditional philosophy of science and social theory assimilate historical episodes to general schemes. History becomes a set of instances of the operation (or failure to operate) of an enduring scientific rationality, or of the working out of enduring social interests, or whatever. Traditional interpretations thus tend to draw us away from historical and cultural specificity and toward an invariant underlying humanist skeleton; and history thus tends to appear as an endless repetition of the same (as I think Donna Haraway puts it). The past becomes a litany of, say, endless, and endlessly depressing, clashes within a standard array of interests. We arrive at an image of a world in which, to recycle an earlier quotation from David Noble, "everything changes, yet nothing moves." History as macromangling, however, looks somewhat different. It is not, of course, that my analysis refuses to find a pattern in history. Most of this book has been based on historical instances offered as exemplifications of the pattern that I call the mangle. But inasmuch as the mangle lacks the substantive explanatory variables of tradi-

38. On the history of the industrial research laboratory, see Noble 1979; van den Belt and Rip 1987; and Hounshell 1992. Van den Belt and Rip's essay includes fascinating detail on the mangling of patent law around the industrialization of organic chemistry research in Germany in the late nineteenth century. Noble shows that many of the features I noted in the mangling of science and the military in World War II can also be identified in prior manglings of science and industry: the MIT Technology Plan established in 1920, for example, is an early example of double panopticism: a tactic intended to make the interior of the research university visible to industry, and vice versa (Noble 1979, 142–44).

tional theory—epistemic rules, enduring human actors and their interests, constraints and limits, and so forth—so it makes possible a kind of double vision, both delineating historical processes in a form that we can grasp and, at the same time, continually throwing us back upon cultural specificity. Thus my World War II example both assimilates the intersection of science and the military to my overall scheme, and, in the same movement, thematizes very specific transformations in the material, social, and conceptual worlds, and in how they hung together—transformations that have been, as it happens, immensely consequential for the present. One cannot come to terms with how the postwar world differs from all that had gone before without appreciating that "science" and "the military" emerged from World War II as quite different actors from those that entered it, and, unlike traditional explanatory schemes, the mangle *encourages us to notice the difference.*

This, then, is the sense in which, in contrast to an endless history of the same, the mangle can help us to get to grips with the singularity of historical eras, including our own. For me, the promise of a mangle-ish performative historiography of STS is precisely that it offers us a route to comprehending what Lyotard once called our postmodern condition. In the future, I hope to help redeem this promise (Pickering 1994b is a start).

7.5 POSTSCRIPT: NONSTANDARD AGENCY

The preceding sentence was my exit line, but two wilder lines of thought continue to spin through my mind, and hence two postscripts. In this first one, I want to question some ideas about human and nonhuman agency that I have so far taken for granted. I make them explicit and then discuss some situations where they appear not to apply.

Throughout the book, I have confidently distinguished between human and nonhuman agency, between people and machines. And I think I have been right to do so. I have been writing about people who live in worlds where the two are clearly distinguishable, almost by definition. The distinction is there, in practice. Material agency is captured by machines as material objects, separate from us as creatures of flesh and blood. Machines display regular, predictable, and nonvolitional powers that we can set in motion and direct, but that are not reducible to human powers. Further, these machinic powers have expanded fabulously over the centuries, and it is hard to imagine them ever reaching a limit. In contrast, human agency seems intrinsically bounded. In making, setting in motion, and tending each generation of machines, we humans seem to display the same old powers, just disciplined differently. None of the studies I have discussed involve much more of human beings than observing what is going on and manipulating medium-sized objects in humdrum ways—though nowadays those objects are often themselves sophisticated instruments and machines. If, therefore, the human race now has at its disposal more powers than it once had, they are the powers of machines that we have constructed and learned to use; they are not distinctly human powers, proper to human bodies and minds. This is what I shall call the standard view of agency, human and nonhuman.

If we look across the spatial or temporal horizons of modern industrial society, however, we find a different situation. In some cultures of preindustrial Europe, say, we find nonstandard material agency, agency that seems to lack the regularity that we associate with machines and that displays distinctively human characteristics. We find stories, for example, of mines inhabited by dwarves, demons, *cobalos, virinculi montani, Bergmännlein*—volitional entities with whom human beings can deal profitably in the extraction of minerals, but who are liable to transmute precious metals into dirt if their rights are infringed (Webster 1982). On the other side, we also find human beings with quite nonstandard powers—magi, alchemists, witches, and so on (see, for instance, Hannaway 1975). The most vivid illustrations that I know of such powers come not from European history but from Carlos Castaneda's popular anthropology (Castaneda 1968 and its many sequels). One might worry about Castaneda's credentials, but, as discussed below, there are reasons for worrying about any account of nonstandard agency, so let me just rehearse some of Castaneda's claims as an example of what I have in mind.

Castaneda tells us about his partial induction into a "Yaqui way of knowledge," via his apprenticeship to a Yaqui Indian known as Don Juan. Don Juan, says Castaneda, introduced him to a set of complex disciplines through which he was able to emulate some of his master's feats: viewing contemporary events far removed in space, flying like a bird, being in two places at once, conversing with the spirits of hallucinogenic mushrooms, and so on. Such performances sound far-fetched, but Castaneda's writing at least makes them thinkable, and there are reports of similarly far-fetched human abilities in many nonscientific cultures: think of the feats of yogic masters for example—levitation, the suspension of bodily processes, and so forth. These are all examples of what I would call nonstandard human agency; they all entail human performativity that goes beyond that exemplified in the remainder of this book. They are not the kinds of things that "we" do.

So, beyond the sphere of science and technology, we find reports, at least, of nonstandard human and material agency—human and material performances that violate our customary understandings of our own powers and those of the world. Furthermore, we can note that, although I have introduced them separately, such nonstandard agencies seem to transgress the separation of the human and the nonhuman that we take for granted. If miners have to deal with *cobalos,* which are in some sense volitional living beings like us, then, for them, the material and human worlds are immediately and inextricably entangled, right on the surface. Castaneda learned how to augment his material powers from within himself, in a way that we tend to think possible only via the use of machines. Nonstandard agency, then, resides in nondualist—explicitly posthumanist—cultures where our distinctions between the human and nonhuman are eroded, if not entirely effaced. The question arises, of course, of whether we should take accounts of nonstandard agency seriously, and I will come back to this in a moment. First, though, I do want to take such accounts seriously, and I want to ask how they bear upon my present analysis.

My first observation is that I can see no reason why my general analysis of practice should not continue to apply to realms of nonstandard agency. Castaneda, in fact, describes his apprenticeship to Don Juan as, in my terms, a dialectic of resistance and accommodation in a field of agency, through which he eventually began to move away from his dualistic understanding of the world and into the realm of nonstandard performativity. The manglings in question, moreover, involved the production of alignments between multiple and heterogeneous elements—lizards, herbs, fungi, rituals, as well as the specific performances of Castaneda, Don Juan, and others. As Castaneda describes it, he learned to emulate Don Juan much as Sibum learned to emulate Joule (section 3.5). The implication of the examples just mentioned, therefore, is that we need to see agency as itself at stake in practice in a much more radical way than the previous chapters have suggested. To focus, for instance, on the human side of the human/nonhuman couple, it is not simply the case that a circumscribed domain of human performativity can be disciplined in an indefinite number of ways in constituting different cultural configurations. The very envelope of human performativity—what human agency can do, beyond mundane observation and manipulation of material objects—needs itself to be seen as in the plane of practice and subject to mangling.

Next we can note that these remarks on nonstandard agency point toward a very radical incommensurability between our scientific-technological-industrial world and the others just mentioned. When I talked in section 6.2 about the new sense of incommensurability that accompanies the move to the performative idiom and the recognition of shifts in our machinic grip on the world, my examples remained within the orbit of standard conceptions of agency, of machines and people. The present discussion goes beyond that sense of incommensurability to delineate the possibility of a radical *performative incommensurability*—an incommensurability of *powers*—by suggesting the idea that within different cultures human beings and the material world might exhibit capacities for action quite different from those we customarily attribute to them. That our own powers as human beings might be bound up with culture in this way is a quite startling idea that I find it fascinating to dwell upon. It is, of course, absent from traditional discussions of incommensurability; only the shift to the performative idiom makes it even thinkable.

But still, the question remains of whether we should think it. Should we take accounts of nonstandard agency at face value the way that I have taken accounts of standard human and material agency at face value in the preceding chapters? Are not the former delusions, innocent or otherwise? I have no answers to these questions. I feel happy enough talking about standard material and human agency because I have a lot of experience of it, which I presume is shared by my readers. I live in a world of machines having orthodox powers, and I need only credit the scientists I have discussed with human powers similar to my own in tinkering with equipment. In contrast, I have no experience of nonstandard agency, and I doubt whether many readers have either. I can only add one remark concerning the difficulties of arriving at a position on performative incommensurability.

Traditional discussions of incommensurability make much ado about problems of communication, translation, and understanding (chap. 6, n. 11). It is impossible, goes the argument, to get the hang of alien representational systems from within one's own. This seems to me to be a mistake. I think I pretty much have the hang of what Castaneda is talking about, just as I have the hang of the old physics of elementary particles. What I do not know is whether to believe Castaneda. That hinges upon whether human beings *can* acquire the powers that he describes, and to know that would require trying to acquire such powers and succeeding or failing. The problem here is thus not so much one of communication but of performance. It would be nice if one could, as it were, know all that one wants to know about other cultures by reading about them—but this route is, in principle, unavailable in cases of (putative) performative incommensurability.

That is as far as I can get with this line of thought. Incommensurability can indeed be associated with problems of access to alien cultures, but I think that these problems are located at the level of performance, not just that of representation. There is no way out of this. One can only contemplate nonstandard performativity—with fascination or disgust—and perhaps try to learn the odd nonstandard performance oneself (Feyerabend 1978). The performative idiom does, at least, help us to get clear that this last is an option, no matter how much representationalism tries to obscure it.[39]

7.6 POSTSCRIPT: THE TOE MANGLE

> Uniformities are precisely the sort of facts that need to be accounted for . . . Law is *par excellence* the thing that wants a reason. Now the only possible way of accounting for the laws of nature and for uniformity in general is to suppose them results of evolution. This supposes them not to be absolute, not to be obeyed precisely. It makes an element of indeterminacy, spontaneity, or absolute chance in nature . . . [W]e must suppose . . . minute discrepancies to exist owing to the imperfect cogency of the law itself, to a certain swerving of the facts from any definite formula.
>
> Charles Sanders Peirce, "The Architecture of Theories"

Multidisciplinary eclecticism in science studies is often self-described as "evolutionary epistemology" (Campbell 1974, 1993; Giere 1988; Hull 1988; Richards 1987). Analogies are invoked to mechanisms of natural selection in evolutionary biology: instead of biological populations, we are asked to think about populations of ideas (theories, representations) randomly mutating in a selection environment usually conceived as the sum of facts, rules of scientific method (or naturalized substitutes), and social interests. The mangle can also be seen as an evolutionary model

39. One does not necessarily have to travel far from the world of late twentieth-century science to encounter realms of nonstandard agency. See Collins 1992, chap. 5, and Collins and Pinch 1982 on paranormal phenomena.

of science, and it is, I would say, a more interesting and far-reaching one than those just mentioned. On the one hand, it is not just an epistemology. It is about how ideas and so forth evolve, but it is also about the evolution of machines and instruments, human disciplines, and social relations. On the other hand, from the perspective of the mangle, the selection environment is not given in advance (as rules, interests, or whatever), but itself emerges within the evolutionary process—in uncertain captures of material or disciplinary agency and achievements of interactive stabilization.

This much is clear from the emergent posthumanism of the mangle, but now I want to generalize this line of thought in a particular direction. The mangle is not so much modelled upon biological understandings of evolution; I suggest that it is, in its own right, an evolutionary theory of indefinite scope.[40] Though I have so far developed my analysis with specific reference to human practice, that reference can easily be deleted. If we replace my analysis of the intentional structure of human agency with a less structured notion like "drift," and if we relax my determined focus on literal machines, we are left with a schema that might describe the evolution of any field of agency or agencies, nonhuman as well as human.[41]

I cannot see, for example, why we should not think of biological organisms, populations, and species as loci of agency, engaged in open-ended struggles with the agency of other organisms and the inorganic world that are just like the struggles already described with respect to scientists. One could conceptualize all sorts of fascinating evolutionary phenomena, including complex evolutionary interdependences of heterogeneous populations, in terms of dialectics of resistance and accommodation, interactive stabilizations, topological reconfigurations, and so on. The mangle could thus be readily understood as itself a model of biological evolution, though not a neo-Darwinian or sociobiological one. It would, instead, be an irreductive model (Latour 1988a, part 2). Nothing substantive in the biological world would be seen as enduring unchanged and controlling evolution—it would be a model of evolution

40. Biographically, I started thinking about evolutionary biology only after I had worked out my analysis of the mangle. I thank Gerry Geison for helpful discussions on different schools of thought in evolutionary biology and their relation to the present argument.

41. "Drift" is Maturana and Varela's (1992) name for a tendency of the form and properties of any organism to change over time. I thank Barbara Herrnstein Smith for drawing my attention to this book.

without DNA or genes or nonemergent mechanisms of selection of whatever sort.[42] Furthermore, there is no need to stick to biological evolution. One can imagine trying to conceptualize the evolution of the cosmos as a whole—of inorganic as well as organic matter—as evolving within fields of agency in dialectics of resistance and accommodation. One could envisage, say, a physics without quarks, as well as a biology without genes (back to this below).

And so, in the end I allow myself to be overtaken by hubris in thinking of my analysis of scientific practice as a potential TOE, a theory of everything. I am, of course, not alone in yielding to this temptation. Physicists speak of little else these days, and I have offered an analysis of physics. More relevantly, perhaps, almost everyone who recognizes temporal emergence in human practice seems to end up flirting with the idea that they are speaking not only about how the nonhuman world necessarily strikes us but about how it actually is. Such speculation has certainly been endemic to pragmatist thought since its earliest days—witness my opening quotation from Peirce.[43] Where it might lead—whether to the Department of Emergency Studies as the ultimate locus of antidisciplinary synthesis, or to the emergency ward—is not clear to me. I find it challenging, though, and I close with a few remarks that might be relevant.

42. To be more exact, the idea would be that if one finds it productive to think about DNA and genes in reproduction, then one should think of them as being in the plane of practice—the plane of biological production and reproduction—rather than as detached controllers of the process. Maturana and Varela have a similar idea: "That modifications of those components called genes dramatically affect the structure [of cells] is very certain. The error lies in confusing essential participation with unique responsibility. By the same token one could say that the political constitution of a country determines its history. This is obviously absurd. The political constitution is an essential component in any history but it does not contain the 'information' that specifies that history" (1992, 69). To cut off an eclectic—partially emergent, partially nonemergent—reading of this passage, I would note that political constitutions are themselves revisable in practice and that, indeed, history is quite capable of proceeding in their absence.

43. Or see William James: "*for rationalism reality is ready-made and complete from all eternity, while for pragmatism it is still in the making, and awaits part of its complexion from the future.* On the one side the universe is absolutely secure, on the other it is still pursuing its adventures" ([1907, 1909] 1978, 123). Likewise one can read Deleuze and Guattari 1987 in its entirety as a TOE. Popper offers a general observation: "It has often been suggested that instead of vainly attempting to follow in sociology the example of physics, it would be better to follow in physics the example of a historicist sociology . . . Historicists who are anxious to emphasize the unity of physics and sociology are especially inclined to think on such lines" ([1957] 1986, 103 n. 1). He cites an essay by Neurath. I was alerted to the aspect of Peirce's thought discussed here (and to the essay by Cocconi

First, it is worth noting that, while a slippery slope always leads from historicizing understandings of knowledge to TOEs, the TOE mangle is better developed than most of its predecessors.[44] Most conspicuously, it incorporates and exemplifies from the start ideas about nonhuman agency. It is thus directly *about* everything, in contrast to humanist philosophies like classical pragmatism that focus more or less exclusively on human agency and that therefore tend to end up visualizing matter as "effete mind" (Peirce [1891] 1923, 170)—whatever that might mean.[45] At a more detailed level, the mangle offers us a relatively fine-grained analysis of temporal maneuvers in fields of multiple and heterogenous agency with which to try to grasp the evolution of everything.

Of course, the stumbling blocks in the way of any irreductionist and historicizing TOE are the traditional reductionist and nonemergent sciences like molecular biology or particle physics, with their enduring fundamental entities—DNA and quarks—and I want to close by mentioning a couple of potential ways around this resistance. One would be to notice that such nonemergent accounts of nature no longer exert quite the hold they once did over the scientific imagination. Even many physicists are now interested in chaos theory, nonlinear systems, and what have you, in the evolution of complex physical systems whose history is important, and that have often to be explored in their temporal evolution with the aid of simulations and supercomputers. Drawing back somewhat from the claim to have a theory of quite everything, the TOE mangle could clearly ally itself with these latter approaches, and I suspect that inspiration could be drawn from them, even in the specific case of trying to think in general about scientific practice. To lend force to this alliance, one could further observe that traditional nonemergent sciences, for all their intricacy, are highly circumscribed in their empirical grounding. Molecular biology is more about what happens in test

cited in n. 49 below) by reading Schweber forthcoming. I thank him for sending me a prepublication copy.

44. Peirce concludes the essay from which I quoted earlier with the frustrating assertion that his TOE "has been worked out by me with elaboration. It accounts for the main features of the universe as we know it,—the characters of time, space, matter, force, gravitation, electricity, etc. It predicts many more things which new observations can alone bring to the test. May some future student go over this ground again, and have leisure to give his results to the world" ([1891] 1923, 177–78).

45. I thank Bruno Latour for emphasizing the humanism of classical pragmatism to me (and see Latour 1992b, 136, where he speaks of his analysis as "a form of pragmatism, but extended to nonhuman actors"). The quotation from James in note 43 above comes from an essay significantly entitled "Pragmatism and Humanism."

tubes than in families; quark-gauge theory (the "standard model") is not only confined to the particle-physics laboratory (and perhaps the Big Bang) but, as noted in section 6.2, engages only with the rarest phenomena to be observed there. The historicizing TOE mangle—like existing historicizing approaches in the natural sciences—is, in contrast, about the visible, about the world of appearance.[46] The domain of the traditional sciences that the TOE mangle cannot reach is a very small one.[47] But this is too quiet a note to finish on.

The standard riposte to relativist accounts of science, never mind their elevation into TOEs, is to challenge their author to come up with a different science—a different physics, for instance, when the riposte is addressed to people like me who want to write about that field. Show us a physics without quarks! This is a tall order, given the fantastic investment of effort, hardware, and money already invested in the status quo. However, the mangle suggests a way to rise to the bait, a way, in this case, to speak directly to the concerns of traditional particle physics—to what was, until recently, the vanguard of nonemergent reductionist thought. What follows has to be both brief and somewhat technical, but I would like to end with a constructive proposal.

In the 1960s, two very different understandings of the world vied for the soul of particle physics. In one corner was the quark model—a picture of the world as built from the enduring entities we talked about in chapter 3. What physicists had to do, on this view, was determine the properties of quarks and then explain the rest of the phenomenal world on that basis. In the other corner was the so-called S-matrix bootstrap. In that approach, there were no privileged constituents of matter (it was often referred to as a "democratic" understanding of matter, in contrast to the "aristocracy" of quarks). Instead, the S-matrix postulated a set of coupled nonlinear equations interrelating the properties of elementary particles (their masses, charges, etc.) and the forces that existed between them. What physicists had to do, from this perspective, was find some self-consistent solution to these equations in which particles (their masses, charges, etc.) and forces would pull themselves up by their boot-

46. I have argued already against traditional attempts to find enduring substrata in the explanation of human practice; here one has to imagine this argument carried through to the evolution of everything.

47. The complaint about the narrowness of traditional natural science is nicely expressed by Giora Hon (1993), who goes on to call for a new physics of the visible. I thank him for a stimulating discussion on this topic, and I hope that the proposal that follows meets with his approval.

straps, as it were, in relations of reciprocal determination.[48] As it happens, the quark model prospered at the heart of the new physics (section 6.2) and dominates physicists' current TOEs, while the S-matrix bootstrap, as part of the old physics, foundered. I want to suggest, however, that it is possible to resurrect the latter as part of my own TOE.[49]

The bootstrap was itself a mangle-ish approach to particle theory. The idea that particle properties and the forces between them should mutually determine one another via a self-consistency requirement *is* a description of interactive stabilization in the material world. What the bootstrap in its standard version lacks, however, is any place for time. The traditional attitude to the bootstrap equations has been that they must have a unique and atemporal solution, which would specify (correctly) the properties of particles and forces. But this is the line of physical thought that failed. The coupled nonlinear equations were analytically intractable, and no mathematical solution to them could be found, while approximation schemes that made them soluble by truncating them led nowhere very interesting. My suggestion is that one might revivify the S-matrix bootstrap by putting time back in. Perhaps the universe itself has never quite solved the S-matrix equations. Perhaps—this is my general idea—the cosmos is continually out of whack at whatever level we care to examine it, even the tiniest distance scales we can imagine.[50] And perhaps the microworld is continually evolving in response to such mismatches. This suggests, to me at least, that instead of trying to find eternally stable solutions to the S-matrix equations, one might instead try iterating them, as a way of modelling the temporal *evolution* of particles and forces.

Certainly, a lot of interesting techniques for examining the evolution

48. To return to my earlier remarks on the narrowness of elementary-particle theory, it is worth noting that the current exclusive interest in very rare elementary-particle events hangs together with developments in the quark model over the past twenty years or so. In contrast, the S-matrix approach traditionally focused on common phenomena. It is also worth remarking that in the late 1970s quarklike structures were claimed to emerge from S-matrix calculations (Pickering 1984b, 415 n. 13).

49. Pickering 1984b contains much detail on the quark model; section 3.4 discusses the S-matrix approach. The latter has always been more suggestive of outlandish metaphysical speculation than the former: see, for example, Capra 1975. The ideas that follow struck me while reading a remarkable essay by G. Cocconi (1970), a practicing elementary-particle physicist. Cocconi, too, suggests the possibility of an evolutionary understanding of the world of elementary particles, though not along the lines laid out below.

50. This is a TOE generalization of the ideas about scale invariance introduced in section 7.4.

of nonlinear systems have been developed since the heyday of the S-matrix, and it seems to me that interesting findings might well emerge from this approach. If the project were practicable, its upshot would certainly be a historicized and evolutionary vision of the fundamental constituents of matter—though obviously there are no advance guarantees of how persuasive that picture would be. One might hope, for example, to find points of limited stability in the specification of the material world, temporary interactive stabilizations within the multiplicity of open-endedly evolving particles and forces—some of which might relate to the world we presently find ourselves in—rather than final solutions. In the end, who knows what one would find?

This is speculation, and not just hubris but traces of my previous incarnation as a particle physicist are surfacing here. But still, I hope these last two paragraphs are enough to suggest the possibility of an alternative approach to the topics of microphysics that would deserve to be a part of the TOE mangle—that could contribute to (not found) an overall emergent and posthumanist vision of mangle-ish human practice as happening in one corner of a world that is itself a mangle-ish place, a vision in which everything becomes in relation to everything else and nothing is fixed. It is a nice picture to meditate upon—the dance of agency as the dance of Shiva . . .

References

Ackermann, R. J. 1985. *Data, Instruments, and Theory: A Dialectical Approach to Understanding Science.* Princeton: Princeton University Press.

———. 1991. Allan Franklin, Right or Wrong? In A. Fine, M. Forbes, and L. Wessels, eds., *PSA 1990: Proceedings of the 1990 Biennial Meeting of the Philosophy of Science Association,* vol. 2, *Symposium and Invited Papers* (East Lansing, MI: Philosophy of Science Association), pp. 451–57.

Adams, H. 1961. *The Education of Henry Adams: An Autobiography.* Introduction by D. W. Brogan. Boston: Houghton Mifflin.

Alvarez, L. 1987a. *Alvarez: Adventures of a Physicist.* New York: Basic Books.

———. 1987b. Recent Developments in Particle Physics. Nobel lecture. 1968. Reprinted in Trower 1987, pp. 110–53.

Ashmore, M. 1989. *The Reflexive Thesis: Wrighting Sociology of Knowledge.* Chicago: University of Chicago Press.

———. 1993. Behaviour Modification of a Catflap: A Contribution to the Sociology of Things. *Kennis en Methode* 17:214–29.

Ashmore, M., R. Wooffitt, and S. Harding, eds. 1994. *Humans and Others: The Concept of "Agency" and Its Attribution.* Special issue of *American Behavioral Scientist* 37 (6).

Baigrie, B. 1994. Scientific Practice: The View from the Tabletop. In J. Buchwald, ed., *Scientific Practice: Theories and Stories of Physics* (Chicago: University of Chicago Press, 1995), pp. 87–122.

Baird, D. 1993. Analytical Chemistry and the "Big" Scientific Instrumentation Revolution. *Annals of Science* 50:267–90.

Baird, D., and A. Nordmann. 1994. Fact-Well-Put. *British Journal for the Philosophy of Science* 45:37–77.

Barnes, B. 1974. *Scientific Knowledge and Sociological Theory.* London and Boston: Routledge and Kegan Paul.

———. 1977. *Interests and the Growth of Knowledge.* London and Boston: Routledge and Kegan Paul.

———. 1982. *T. S. Kuhn and Social Science.* London: Macmillan.

———. 1991. How Not to Do the Sociology of Knowledge. In A. Megill, ed., *Rethinking Objectivity,* part 1. Special issue of *Annals of Scholarship* 8, nos. 3–4: 321–35.

Barnes, B., and D. Bloor. 1982. Relativism, Rationalism, and the Sociology of Knowledge. In M. Hollis and S. Lukes, eds., *Rationality and Relativism* (Cambridge: MIT Press), pp. 21–47.

Bastide, F. 1990. The Iconography of Scientific Texts: Principles of Analysis. In M. Lynch and S. Woolgar, eds., *Representation in Scientific Practice* (Cambridge: MIT Press), pp. 187–229.

Baudrillard, J. 1988a. *America.* New York and London: Verso.

———. 1988b. *Jean Baudrillard: Selected Writings.* Edited by M. Poster. Stanford: Stanford University Press.

Becchi, C., G. Gallinaro, and G. Morpurgo. 1965. Measurement of Small Charges in Macroscopic Amounts of Matter: Discussion of a Proposed Experiment and Description of Some Preliminary Observations. *Nuovo Cimento* 39:409–12.

Becchi, C., and G. Morpurgo. 1965a. Test of the Nonrelativistic Quark Model for "Elementary" Particles: Radiative Decays of Vector Mesons. *Physical Review* 140B:687–90.

———. 1965b. Vanishing of the E2 Part of the $N^*_{33} \rightarrow N\text{-}\gamma$ Amplitude in the Non-relativistic Quark Model of "Elementary" Particles. *Physics Letters* 17:352–54.

Bernal, J. D. 1953a. *Science and Industry in the Nineteenth Century.* London: Routledge and Kegan Paul.

———. 1953b. Science, Industry, and Society in the Nineteenth Century. In S. Lilley, ed., *Essays on the Social History of Science,* special issue of *Centaurus* 3:138–65.

———. 1969. *Science in History.* 4 vols. 3d ed. Cambridge: MIT Press. 1st ed. published 1954.

Bernstein, R. 1983. *Beyond Objectivism and Relativism: Science, Hermeneutics, and Practice.* Philadelphia: University of Pennsylvania Press.

Beyerchen, A. D. 1989. Nonlinear Science and the Unfolding of a New Intellectual Vision. *Papers in Comparative Studies* 6:25–49.

Bhaskar, R. 1975. *A Realist Theory of Science.* Leeds: Leeds Books.

Biagioli, M. 1990a. The Anthropology of Incommensurability. *Studies in History and Philosophy of Science* 21:183–209.

———. 1990b. Galileo the Emblem Maker. *Isis* 81:230–58.

Bijker, W., T. Hughes, and T. J. Pinch, eds. 1987. *The Social Construction of Technological Systems: New Directions in the Sociology and History of Technology.* Cambridge: MIT Press.

Bijker, W., and J. Law, eds. 1992. *Shaping Technology/Building Society: Studies in Sociotechnical Change.* Cambridge: MIT Press.

Bloor, D. 1981. Hamilton and Peacock on the Essence of Algebra. In H. Mehrtens, H. Bos, and I. Schneider, eds., *Social History of Nineteenth Century Mathematics* (Boston: Birkhäuser), pp. 202–32.

———. 1983. *Wittgenstein: A Social Theory of Knowledge.* London: Macmillan.

———. 1991. *Knowledge and Social Imagery.* 2d ed. Chicago: University of Chicago Press. 1st ed. published 1976.

———. 1992. Left and Right Wittgensteinians. In A. Pickering, ed., *Science as Practice and Culture* (Chicago and London: University of Chicago Press), pp. 266–82.

Boole, G. 1847. *Mathematical Analysis of Logic: Being an Essay towards a Calculus of Deductive Reasoning.* Cambridge: Macmillan, Barclay, and Macmillan.

Braverman, H. 1974. *Labor and Monopoly Capital: The Degradation of Work in the Twentieth Century.* New York: Monthly Review Press.

Brown, J. L., D. A. Glaser, and M. L. Perl. 1956. Liquid Xenon Bubble Chamber. *Physical Review* 102:586–87.

Bugg, D. V. 1959. The Bubble Chamber. *Progress in Nuclear Physics* 7:1–52.

Callon, M. 1986. Some Elements of a Sociology of Translation: Domestication of the Scallops and the Fishermen of St Brieuc Bay. In J. Law, ed., *Power, Action and Belief: A New Sociology of Knowledge?* Sociological Review Monograph 32 (London: Routledge and Kegan Paul), pp. 196–233.

———. 1987. Society in the Making: The Study of Technology as a Tool for Sociological Analysis. In W. Bijker, T. Hughes, and T. J. Pinch, eds., *The Social Construction of Technological Systems: New Directions in the Sociology and History of Technology* (Cambridge: MIT Press), pp. 83–103.

———. 1991. Techno-Economic Networks and Irreversibility. In J. Law, ed., *A Sociology of Monsters? Essays on Power, Technology, and Domination* (London: Routledge), pp. 132–61.

———. 1994. Four Models for the Dynamics of Science. In S. Jasanoff, G. E. Markle, J. C. Petersen, and T. J. Pinch, eds., *Handbook of Science and Technology Studies.* Los Angeles: Sage.

Callon, M., and B. Latour. 1981. Unscrewing the Big Leviathan, or How Do Actors Macrostructure Reality? In K. Knorr-Cetina and A. Cicourel, eds., *Advances in Social Theory and Methodology: Toward an Integration of Micro- and Macro-Sociologies* (Boston: Routledge and Kegan Paul), pp. 277–303.

———. 1992. Don't Throw the Baby Out with the Bath School! A Reply to Collins and Yearley. In A. Pickering, ed., *Science as Practice and Culture* (Chicago and London: University of Chicago Press), pp. 343–68.

Callon, M., and J. Law. 1989. On the Construction of Sociotechnical Networks: Content and Context Revisited. In L. Hargens, R. A. Jones, and A. Pickering, eds., *Knowledge and Society: Studies in the Sociology of Science, Past and Present,* vol. 8 (Greenwich, CT: JAI Press), pp. 57–83.

Campbell, D. T. 1974. Evolutionary Epistemology. In P. A. Schilpp, ed., *The Philosophy of Karl R. Popper,* Library of Living Philosophers, vol. 14, part 1 (LaSalle, IL: Open Court), pp. 413–63. Reprinted in *Methodology and Epistemology for Social Science,* edited by E. S. Overman (Chicago: University of Chicago Press, 1988), pp. 393–434.

———. 1993. Plausible Coselection of Belief by Referent: All the "Objectivity" That Is Possible. *Perspectives on Science* 1:88–108.

Capra, F. 1975. *The Tao of Physics.* London: Fontana.

Cartwright, N. 1983. *How the Laws of Physics Lie.* Oxford: Oxford University Press.

———. 1989. *Nature's Capacities and Their Measurement.* Oxford: Clarendon Press.

Castaneda, C. 1968. *The Teachings of Don Juan: A Yaqui Way of Knowledge.* Harmondsworth, England: Penguin.

Chalmers, A. 1992. Is a Law Reasonable to a Hume? *Cogito* (winter): 125–29.

Christie, J. R. R. 1993. Aurora, Nemesis, and Clio. In J. A. Secord, ed., *The Big Picture,* special issue of *British Journal for History of Science* 36 (4): 391–405.

Clarke, A., and J. Fujimura, eds. 1992. *The Right Tool for the Job: At Work in Twentieth Century Life Science.* Princeton: Princeton University Press.

Cocconi, G. 1970. The Role of Complexity in Nature. In M. Conversi, ed., *Evolution of Particle Physics* (New York: Academic Press), pp. 81–87.

Cohen, M. R. 1923. Introduction to C. S. Peirce, *Chance, Love, and Logic: Philosophical Essays,* edited by Morris R. Cohen. New York: Harcourt, Brace.

Collini, S. 1994. Escape from DWEMsville. *Times Literary Supplement,* 27 May 1994, pp. 3–4.

Collins, H. M. 1990. *Artificial Experts: Social Knowledge and Intelligent Machines.* Cambridge: MIT Press.

———. 1992. *Changing Order: Replication and Induction in Scientific Practice.* 2d ed. Chicago: University of Chicago Press. 1st ed. published 1985.

———. 1994. Dissecting Surgery: Forms of Life Depersonalized. *Social Studies of Science* 24:311–33.

Collins, H. M., and M. Kusch. Forthcoming a. Automating Airpumps: An Empirical and Conceptual Analysis.

———. Forthcoming b. Two Kinds of Actions: A Phenomenological Study.

Collins, H. M., and T. J. Pinch. 1982. *Frames of Meaning: The Social Construction of Extraordinary Science.* London: Routledge and Kegan Paul.

Collins, H. M., and S. Yearley. 1992a. Epistemological Chicken. In A. Pickering, ed., *Science as Practice and Culture* (Chicago and London: University of Chicago Press), pp. 301–26.

———. 1992b. Journey into Space. In A. Pickering, ed., *Science as Practice and Culture* (Chicago and London: University of Chicago Press), pp. 369–89.

Crowe, M. J. 1985. *A History of Vector Analysis: The Evolution of the Idea of a Vectorial System.* New York: Dover.

———. 1988. Ten Misconceptions about Mathematics and Its History. In W. Aspray and P. Kitcher, eds., *History and Philosophy of Modern Mathematics* (Minneapolis: University of Minnesota Press), pp. 260–77.

———. 1990. Duhem and History and Philosophy of Mathematics. *Synthese* 83:431–47.

Cultural Studies. 1994. *Times Literary Supplement,* 27 May 1994, pp. 3–13.

Cunningham, A., and P. Williams. 1993. De-Centring the "Big Picture": *The Origins of Modern Science* and the Modern Origins of Science. *British Journal for the History of Science* 26:407–32.

Cussins, A. 1992a. Constructing a World in a Painting. In A. Cussins, B. Latour,

A. Lowe, and B. Smith, *Registration Marks: Metaphors for Subobjectivity* (London: Pomeroy Purdy Gallery), pp. 9–21.

———. 1992b. Content, Embodiment, and Objectivity: The Theory of Cognitive Trails. *Mind* 101:651–88.

Daston, L. 1993. The Moralized Objectivities of Nineteenth-Century Science. University of Chicago. Manuscript.

Daston, L., and P. Galison. 1992. The Image of Objectivity. *Representations* 40:81–128.

Davidson, D. 1973–74. On the Very Idea of a Conceptual Scheme. *Proceedings of the American Philosophical Association* 17:5–20.

Deleuze, G., and F. Guattari. 1987. *A Thousand Plateaus: Capitalism and Schizophrenia*. Minneapolis: University of Minnesota Press.

Deleuze, G., and C. Parnet. 1987. *Dialogues*. New York: Columbia University Press.

Denzin, N. 1992. *Symbolic Interactionism and Cultural Studies: The Politics of Interpretation*. Oxford: Blackwell.

Dewey, J. 1923. The Pragmatism of Peirce. In C. S. Peirce, *Chance, Love, and Logic: Philosophical Essays*, edited by M. R. Cohen (New York: Harcourt, Brace), pp. 301–8.

Duhem, P. 1991. *The Aim and Structure of Physical Theory*. Princeton: Princeton University Press.

Edwards, D., M. Ashmore, and J. Potter. 1994. Death and Furniture: The Rhetoric, Politics, and Theology of Bottom Line Arguments against Relativism. *History of the Human Sciences* 7 (4).

Eisenstein, E. L. 1983. *The Printing Revolution in Early Modern Europe*. Cambridge: Cambridge University Press.

Feyerabend, P. K. 1962. Explanation, Reduction, and Empiricism. In H. Feigl and G. Maxwell, eds., *Scientific Explanation, Space, and Time*, Minnesota Studies in the Philosophy of Science, vol. 3 (Minneapolis: University of Minnesota Press), pp. 28–97. Reprinted in *Realism, Rationalism, and Scientific Method: Philosophical Papers* (Cambridge: Cambridge University Press, 1981), 1:44–96.

———. 1965a. On the Meaning of Scientific Terms. *Journal of Philosophy* 62:266–74.

———. 1965b. Problems of Empiricism. In R. Colodny, ed., *Beyond the Edge of Certainty*. Englewood Cliffs, NJ: Prentice Hall.

———. 1975. *Against Method*. London: New Left Books.

———. 1978. *Science in a Free Society*. London: New Left Books.

Fine, A. Forthcoming. Science Made Up: Constructivist Sociology of Scientific Knowledge. In P. Galison and D. Stump, eds., *The Disunity of Science: Boundaries, Contests, and Power*. Stanford: Stanford University Press.

Fisher, H. A. L. 1935. *A History of Europe*. Boston: Houghton Mifflin.

Fleck, L. [1935] 1979. *Genesis and Development of a Scientific Fact*. Chicago: University of Chicago Press.

Fortun, M., and S. S. Schweber. 1993. Scientists and the Legacy of World War II: The Case of Operations Research. *Social Studies of Science* 23:595–642.

Foucault, M. 1972. *The Archaeology of Knowledge*. New York: Pantheon.

———. 1979. *Discipline and Punish: The Birth of the Prison.* New York: Vintage Books.

Franklin, A. 1990. *Experiment, Right or Wrong.* Cambridge: Cambridge University Press.

———. 1991. Do Mutants Have to Be Slain, or Do They Die of Natural Causes? The Case of Atomic Parity Violation Experiments. In A. Fine, M. Forbes, and L. Wessels, eds., *PSA 1990: Proceedings of the 1990 Biennial Meeting of the Philosophy of Science Association,* vol. 2, *Symposium and Invited Papers* (East Lansing, MI: Philosophy of Science Association), pp. 487–94.

Fujimura, J. 1992. Crafting Science: Standardized Packages, Boundary Objects, and "Translation." In A. Pickering, ed., *Science as Practice and Culture* (Chicago and London: University of Chicago Press), pp. 168–211.

Fuller, S. 1992. Social Epistemology and the Research Agenda of Science Studies. In A. Pickering, ed., *Science as Practice and Culture* (Chicago and London: University of Chicago Press), pp. 390–428.

Galison, P. 1983. How the First Neutral Current Experiments Ended. *Reviews of Modern Physics* 55:477–509.

———. 1985. Bubble Chambers and the Experimental Workplace. In P. Achinstein and O. Hannaway, eds., *Observation, Experiment, and Hypothesis in Modern Physical Science* (Cambridge: MIT Press), pp. 309–73.

———. 1987. *How Experiments End.* Chicago: University of Chicago Press.

———. 1988a. Multiple Constraints, Simultaneous Solutions. *PSA 1988: Proceedings of the 1988 Biennial Meeting of the Philosophy of Science Association,* vol. 2, *Symposium and Invited Papers* (East Lansing, MI: Philosophy of Science Association), pp. 157–63.

———. 1988b. Physics between War and Peace. In E. Mendelsohn, M. R. Smith, and P. Weingart, eds., *Science, Technology, and the Military: Sociology of the Sciences Yearbook, 1988* (Dordrecht: Kluwer), pp. 47–86.

———. 1990. Aufbau/Bauhaus: Logical Positivism and Architectural Modernism. *Critical Inquiry* 16:709–52.

———. 1991. Artificial Reality. Paper presented at a conference, "Disunity and Contextualism," Stanford University, 31 March–1 April.

———. 1993. The Ontology of the Enemy. Paper presented to the Berlin Summer Academy on Large Technical Systems, 27 July.

———. 1994. Context and Constraints. In J. Buchwald, ed., *Scientific Practice: Theories and Stories of Physics* (Chicago: University of Chicago Press, 1995), pp. 13–41.

Galison, P., and A. Assmus. 1989. Artificial Clouds, Real Particles. In D. Gooding, T. J. Pinch, and S. Schaffer, eds., *The Uses of Experiment: Studies of Experimentation in the Natural Sciences* (Cambridge: Cambridge University Press), pp. 225–74.

Galison, P., and B. Hevly, eds. 1992. *Big Science: The Growth of Large-Scale Research.* Stanford: Stanford University Press.

Gallinaro, G., M. Marinelli, and G. Morpurgo. 1977. Electric Neutrality of Matter. *Physical Review Letters* 38:1255–58.

Gallinaro, G., and G. Morpurgo. 1966. Preliminary Results in the Search for

Fractionally Charged Particles by the Magnetic Levitation Electrometer. *Physics Letters* 23:609–13.

Garfinkel, H., M. Lynch, and E. Livingston. 1981. The Work of a Discovering Science Construed with Materials from the Optically Discovered Pulsar. *Philosophy of the Social Sciences* 11:131–58.

Geison, G. L. 1993. Research Schools and New Directions in the Historiography of Science. *Osiris* 8:227–38.

Gell-Mann, M. 1964. A Schematic Model of Baryons and Mesons. *Physics Letters* 8:214–15.

Giddens, A. 1984. *The Constitution of Society: Outline of the Theory of Structuration.* Berkeley: University of California Press.

Giere, R. N. 1973. History and Philosophy of Science: Intimate Relationship or Marriage of Convenience? *British Journal for Philosophy of Science* 24:282–97.

———. 1988. *Explaining Science: A Cognitive Approach.* Chicago and London: University of Chicago Press.

———. 1993. Science and Technology Studies: Prospects for an Enlightened Postmodern Synthesis. *Science, Technology, and Human Values* 18:102–12.

Gingras, Y. 1994. Following Scientists through Society? Yes, but at Arm's Length! In J. Buchwald, ed., *Scientific Practice: Theories and Stories of Physics* (Chicago: University of Chicago Press, 1995), pp. 123–48.

Gingras, Y., and S. S. Schweber. 1986. Constraints on Construction. *Social Studies of Science* 16:372–83.

Ginzburg, C. 1980. *The Cheese and the Worms: The Cosmos of a Sixteenth-Century Miller.* New York and London: Penguin.

———. 1989. Clues: Roots of an Evidential Paradigm. In *Myths, Emblems, Clues* (London: Hutchinson Radius), pp. 96–125.

Glaser, D. A. 1958. The Bubble Chamber. *Handbuch der Physik* 45:314–41.

———. 1964. Elementary Particles and Bubble Chambers. Nobel lecture, 1960. Reprinted in *Nobel Lectures: Physics, 1942–1962* (Amsterdam: Elsevier), pp. 529–51.

Gooding, D. 1989. History in the Laboratory: Can We Tell What Really Went On? In F. A. J. L. James, ed., *The Development of the Laboratory: Essays on the Place of Experiment in Industrial Civilization* (London: Macmillan), pp. 63–82.

———. 1990. *Experiment and the Making of Meaning.* Dordrecht: Kluwer Academic.

———. 1992. Putting Agency Back into Observation. In A. Pickering, ed., *Science as Practice and Culture* (Chicago and London: University of Chicago Press), pp. 65–112.

Gooding, D., T. J. Pinch, and S. Schaffer, eds. 1989. *The Uses of Experiment: Studies of Experimentation in the Natural Sciences.* Cambridge: Cambridge University Press.

Graham, L. 1988. Managing Uncertainty in a Graduate Research Practicum. Paper presented at the Midwest Sociological Association Conference, Minneapolis, March.

———. 1994. Critical Biography without Subjects and Objects: An Encounter with Dr. Lillian Moller Gilbreth. *Sociological Quarterly* 35:621–43.

Greenblatt, S. 1991. *Marvelous Possessions: The Wonder of the New World.* Chicago: University of Chicago Press.

Grossberg, L., C. Nelson, and P. Treichler, eds. 1992. *Cultural Studies.* New York: Routledge.

Hacking, I. 1982. Language, Truth, and Reason. In M. Hollis and S. Lukes, eds., *Rationality and Relativism* (Cambridge: MIT Press), pp. 48–66.

———. 1983. *Representing and Intervening.* Cambridge: Cambridge University Press.

———. 1987. Was There a Probabilistic Revolution 1800–1930? In L. Krüger, L. Daston, and M. Heidelberger, eds., *The Probabilistic Revolution,* vol. 1. Cambridge: MIT Press.

———. 1992a. The Self-Vindication of the Laboratory Sciences. In A. Pickering, ed., *Science as Practice and Culture* (Chicago and London: University of Chicago Press), pp. 29–64.

———. 1992b. Statistical Language, Statistical Truth, and Statistical Reason: The Self-Authentification of a Style of Scientific Reasoning. In E. McMullin, ed., *The Social Dimensions of Science* (Notre Dame, IN: University of Notre Dame Press), pp. 130–57.

———. 1992c. "Style" for Historians and Philosophers. *Studies in History and Philosophy of Science* 23:1–20.

Hamilton, W. R. 1837. Theory of Conjugate Functions, or Algebraic Couples, with a Preliminary and Elementary Essay on Algebra as the Science of Pure Time. *Transactions of the Royal Irish Academy* 17:293–422. Reprinted in *The Mathematical Papers of Sir William Rowan Hamilton* (Cambridge: Cambridge University Press), 3:3–96.

———. 1843a. Letter to Graves on Quaternions; or on a New System of Imaginaries in Algebra (17 October 1843). *Philosophical Magazine* 25:489–95. Reprinted in *The Mathematical Papers of Sir William Rowan Hamilton* (Cambridge: Cambridge University Press), 3:106–10.

———. 1843b. Quaternions. Note-book 24.5, entry for 16 October 1843. Reprinted in *The Mathematical Papers of Sir William Rowan Hamilton* (Cambridge: Cambridge University Press), 3:103–5.

———. 1853. Preface to *Lectures on Quaternions* (Dublin: Hodges and Smith). Reprinted in *The Mathematical Papers of Sir William Rowan Hamilton* (Cambridge: Cambridge University Press), 3:117–55.

———. 1967. *The Mathematical Papers of Sir William Rowan Hamilton.* Vol. 3, *Algebra.* Cambridge: Cambridge University Press.

Hankins, T. L. 1980. *Sir William Rowan Hamilton.* Baltimore: Johns Hopkins University Press.

Hannaway, O. 1975. *The Chemists and the Word: The Didactic Origins of Chemistry.* Baltimore and London: Johns Hopkins University Press.

———. 1986. Laboratory Design and the Aim of Science: Andreas Libavius versus Tycho Brahe. *Isis* 77:585–610.

Hanson, N. R. 1958. *Patterns of Discovery: An Inquiry into the Conceptual Foundations of Knowledge.* Cambridge: Cambridge University Press.

Haraway, D. 1979. The Biological Enterprise: Sex, Mind, and Profit from Human Engineering to Sociobiology. *Radical History Review* 20:206–37. Reprinted in *Simians, Cyborgs, and Women: The Reinvention of Nature* (London: Free Association Books, 1991), pp. 43–68.

———. 1985. A Manifesto for Cyborgs: Science, Technology, and Socialist Feminism in the 1980s. *Socialist Review* 80:65–107. Reprinted as "A Cyborg Manifesto: Science, Technology, and Socialist-Feminism in the Late Twentieth Century," in *Simians, Cyborgs, and Women: The Reinvention of Nature* (London: Free Association Books, 1991), pp. 149–81.

———. 1991a. Science as Culture: Science Studies as Cultural Studies? Paper presented at a conference, "Disunity and Contextualism: New Directions in the Philosophy of Science Studies," Stanford University, 31 March–1 April.

———. 1991b. *Simians, Cyborgs, and Women: The Reinvention of Nature.* London: Free Association Books.

———. 1992. The Promises of Monsters: A Regenerative Politics for Inappropiate/d Others. In L. Grossberg, C. Nelson, and P. Treichler, eds., *Cultural Studies* (New York: Routledge), pp. 295–337.

Harré, R., and E. H. Madden. 1975. *Causal Powers: A Theory of Natural Necessity.* Oxford: Blackwell.

Hartley, B. n. d. The Living Academies of Nature: Learning and Communicating New Skills of Early Nineteenth Century Landscape Painting. London. Mimeo.

Heilbron, J. L., and R. W. Seidel. 1990. *Lawrence and His Laboratory: A History of the Lawrence Berkeley Laboratory.* Vol. 1. Berkeley: University of California Press.

Heilbron, J. L., R. W. Seidel, and B. R. Wheaton. 1981. *Lawrence and His Laboratory: Nuclear Science at Berkeley, 1931–1961.* Berkeley: University of California.

Hendry, J. 1984. The Evolution of William Rowan Hamilton's View of Algebra as the Science of Pure Time. *Studies in History and Philosophy of Science* 15:63–81.

Hermann, A., J. Krige, U. Mersits, and D. Pestre. 1987. *History of CERN.* Vol. 1, *Launching the European Organization for Nuclear Research.* Amsterdam: North-Holland.

———. 1990. *History of CERN.* Vol. 2, *Building and Running the Laboratory, 1954–1965.* Amsterdam: North-Holland.

Hesse, M. B. 1966. *Models and Analogies in Science.* Notre Dame, IN: University of Notre Dame Press.

———. 1974. *The Structure of Scientific Inference.* London: Macmillan.

———. 1980. *Revolutions and Reconstructions in the Philosophy of Science.* Brighton, England: Harvester Press.

Hessen, B. 1932. *The Social and Economic Roots of Newton's "Principia." Science at the Crossroads* (London: Kniga), pp. 147–212. Reprint, New York: Howard Fertig, 1971.

Hollis, M., and S. Lukes, eds. 1982. *Rationality and Relativism.* Cambridge: MIT Press.

Holmes, F. L. 1974. *Claude Bernard and Animal Chemistry: The Emergence of a Scientist.* Cambridge: Harvard University Press.

———. 1981. The Fine Structure of Scientific Creativity. *History of Science* 19:60–70.

———. 1985. *Lavoisier and the Chemistry of Life: An Exploration of Scientific Creativity.* Madison and London: University of Wisconsin Press.

Holton, G. 1978. Subelectrons, Presuppositions, and the Millikan-Ehrenhaft Dispute. In *The Scientific Imagination: Case Studies* (Cambridge: Cambridge University Press), pp. 25–83.

Hon, G. 1993. The Unnatural Nature of the Laws of Nature: Symmetry and Asymmetry. In S. French and H. Kamminga, eds., *Correspondence, Invariance, and Heuristics: Essays in Honour of Heinz Post* (Dordrecht: Kluwer), pp. 171–87.

Hopwood, A. 1987. The Archaeology of Accounting Systems. *Accounting, Organizations, and Society* 12:207–34. Reprinted in *Accounting from the Outside: The Collected Papers of Anthony G. Hopwood* (New York: Garland, 1988), pp. 443–70.

Hounshell, D. A. 1992. Du Pont and the Management of Large-Scale Research and Development. In P. Galison and B. Hevly, eds., *Big Science: The Growth of Large-Scale Research* (Stanford: Stanford University Press), pp. 236–61.

Hughes, T. P. 1983. *Networks of Power: Electrification in Western Society.* Baltimore: Johns Hopkins University Press.

Hull, D. L. 1988. *Science as a Process: An Evolutionary Account of the Social and Conceptual Development of Science.* Chicago and London: University of Chicago Press.

Husserl, E. 1970. *The Crisis of European Sciences and Transcendental Phenomenology.* Evanston: Northwestern University Press.

James, W. [1907, 1909] 1978. *Pragmatism and The Meaning of Truth.* Cambridge: Harvard University Press.

Jameson, F. 1984. Postmodernism, or The Cultural Logic of Late Capitalism. *New Left Review* 146 (July–August): 53–94. Reprinted in *Postmodernism, or The Cultural Logic of Late Capitalism* (Durham, NC: Duke University Press, 1991), pp. 1–54.

Jordan, K., and M. Lynch. 1992. The Sociology of a Genetic Engineering Technique: Ritual and Rationality in the Performance of the "Plasmid Prep." In A. Clarke and J. Fujimura, eds., *The Right Tool for the Job: At Work in Twentieth Century Life Science* (Princeton: Princeton University Press), pp. 77–114.

Joule, J. P. 1850. On the Mechanical Equivalent of Heat. *Philosophical Transactions,* part 1, read 21 June 1849. Reprinted in *The Scientific Papers* (London: Taylor and Francis, 1884), 1:298–328.

Keller, E. F. 1985. *Reflections on Gender and Science.* New Haven and London: Yale University Press.

Kellert, S. H. 1993. *In the Wake of Chaos: Unpredictable Order in Dynamical Systems.* Chicago: University of Chicago Press.

Kevles, D. J. 1987. *The Physicists: The History of a Scientific Community in*

Modern America. 2d ed. Cambridge: Harvard University Press. 1st ed. published in 1977.

Kevles, D. J., and G. Geison. 1995. The Experimental Life Sciences in the Twentieth Century: Needs and Opportunities for Historical Research. *Osiris* 11.

Kitcher, P. 1983. *The Nature of Mathematical Knowledge*. Oxford: Oxford University Press.

———. 1988. Mathematical Naturalism. In W. Aspray and P. Kitcher, eds., *History and Philosophy of Modern Mathematics* (Minneapolis: University of Minnesota Press), pp. 293–325.

Knorr-Cetina, K. 1981. *The Manufacture of Knowledge: An Essay on the Constructivist and Contextual Nature of Science*. Oxford and New York: Pergamon.

———. 1988. The Micro-Social Order: Towards a Reconception. In G. Fielding, ed., *Actions and Structure* (London: Sage), pp. 21–53.

———. 1992. The Couch, the Cathedral, and the Laboratory: On the Relation between Experiment and Laboratory in Science. In A. Pickering, ed., *Science as Practice and Culture* (Chicago and London: University of Chicago Press), pp. 113–38.

———. 1994. Laboratory Studies: The Cultural Approach to the Study of Science. In S. Jasanoff, G. E. Markle, J. C. Petersen, and T. J. Pinch, eds., *Handbook of Science and Technology Studies*. Los Angeles: Sage.

Kohler, R. E. 1994. *Lords of the Fly: Drosophila Genetics and the Experimental Life*. Chicago: University of Chicago Press.

Krausz, M., ed. 1989. *Relativism: Interpretation and Confrontation*. Notre Dame, IN: University of Notre Dame Press.

Krieger, M. H. 1992. *Doing Physics: How Physicists Take Hold of the World*. Bloomington: Indiana University Press.

Kuhn, T. 1970. *The Structure of Scientific Revolutions*. 2d ed. Chicago: University of Chicago Press. 1st ed. published 1962.

———. 1983. Commensurability, Comparability, Communicability. In P. D. Asquith and T. Nickles, eds., *PSA 1982: Proceedings of the 1982 Biennial Meeting of the Philosophy of Science Association*, vol. 2, *Symposium and Invited Papers* (East Lansing, MI: Philosophy of Science Association), pp. 669–88.

———. 1991. The Road since Structure. In A. Fine, M. Forbes, and L. Wessels, eds., *PSA 1990: Proceedings of the 1990 Biennial Meeting of the Philosophy of Science Association*, vol. 2, *Symposium and Invited Papers* (East Lansing, MI: Philosophy of Science Association), pp. 3–13.

———. 1992. The Trouble with the Historical Philosophy of Science. Robert and Maurine Rothschild Distinguished Lecture, 19 November 1991. Occasional Publication of the Department of History of Science, Harvard University.

Lakatos, I. 1970. Falsification and the Methodology of Scientific Research Programmes. In I. Lakatos and A. Musgrave, eds., *Criticism and the Growth of Knowledge* (Cambridge: Cambridge University Press), pp. 91–196.

———. 1976. *Proofs and Refutations: The Logic of Mathematical Discovery.* Cambridge: Cambridge University Press.

———. 1978. *Philosophical Papers.* Vol. 1, *The Methodology of Scientific Research Programmes.* Cambridge: Cambridge University Press.

Latour, B. 1983. Give Me a Laboratory and I Will Raise the World. In K. Knorr-Cetina and M. Mulkay, eds., *Science Observed: Perspectives on the Social Study of Science* (Beverly Hills: Sage), pp. 141–70.

———. 1987. *Science in Action: How to Follow Scientists and Engineers through Society.* Cambridge: Harvard University Press.

———. 1988a. *The Pasteurization of France.* Cambridge: Harvard University Press.

———. 1988b. The Politics of Explanation: An Alternative. In S. Woolgar, ed., *Knowledge and Reflexivity: New Frontiers in the Sociology of Knowledge* (Beverly Hills: Sage), pp. 155–76.

———. 1990. Postmodern? No, Simply Amodern! Steps towards an Anthropology of Science. *Studies in History and Philosophy of Science* 21:145–71.

———. 1992a. A "Matter" of Life and Death—Or Should We Avoid Hylozoism? Paper presented at the Department of History and Philosophy of Science, University of Cambridge, 29 October.

———. 1992b. Pasteur on Lactic Acid Yeast: A Partial Semiotic Analysis. *Configurations* 1:129–45.

———. 1993a. On Technical Mediation. Messenger Lectures on the Evolution of Civilization, Cornell University, April 1993.

———. 1993b. *We Have Never Been Modern.* Cambridge: Harvard University Press.

Latour, B., and S. Woolgar. 1986. *Laboratory Life: The Construction of Scientific Facts.* 2d ed. Princeton: Princeton University Press. 1st ed. published 1979.

Laudan, L. 1984a. A Confutation of Convergent Realism. In J. Leplin, ed., *Scientific Realism* (Berkeley: University of California Press), pp. 218–49.

———. 1984b. *Science and Values: The Aims of Science and Their Role in Scientific Debate.* Berkeley: University of California Press.

———. 1987. Progress or Rationality? The Prospects for Normative Naturalism. *American Philosophical Quarterly* 24:19–31.

———. 1989. Thoughts on HPS: 20 Years Later. *Studies in History and Philosophy of Science* 20:9–13.

Law, J. 1987. Technology and Heterogeneous Engineering: The Case of Portuguese Expansion. In W. Bijker, T. Hughes, and T. J. Pinch, eds., *The Social Construction of Technological Systems: New Directions in the Sociology and History of Technology* (Cambridge: MIT Press), pp. 111–34.

———. 1993. *Modernity, Myth, and Materialism.* Oxford: Blackwell.

Law, J., and W. Bijker. 1992. Postscript: Technology, Stability, and Social Theory. In W. Bijker and J. Law, eds., *Shaping Technology/Building Society: Studies in Sociotechnical Change* (Cambridge: MIT Press), pp. 290–308.

Law, J., and A. Mol. 1995. Notes on Materiality and Sociality. *Sociological Review* 43.

Lenoir, T. 1992. Practical Reason and the Construction of Knowledge: The Life-world of Haber-Bosch. In E. McMullin, ed., *The Social Dimensions of Science* (Notre Dame, IN: University of Notre Dame Press), pp. 158–97.

Leplin, J., ed. 1984. *Scientific Realism.* Berkeley: University of California Press.

Linebaugh, P. 1992. *The London Hanged: Crime and Civil Society in the Eighteenth Century.* Cambridge: Cambridge University Press.

Livingston, E. 1986. *The Ethnomethodological Foundations of Mathematics.* Boston and London: Routledge and Kegan Paul.

Lynch, M. 1985a. *Art and Artifact in Laboratory Science: A Study of Shop Work and Shop Talk in a Research Laboratory.* London: Routledge and Kegan Paul.

———. 1985b. Discipline and the Material Form of Images: An Analysis of Scientific Visibility. *Social Studies of Science* 15:37–66.

———. 1991a. Allan Franklin's Transcendental Physics. In A. Fine, M. Forbes, and L. Wessels, eds., *PSA 1990: Proceedings of the 1990 Biennial Meeting of the Philosophy of Science Association,* vol. 2, *Symposium and Invited Papers* (East Lansing, MI: Philosophy of Science Association), pp. 471–85.

———. 1991b. Laboratory Space and the Technological Complex: An Investigation of Topical Contextures. *Science in Context* 4:51–78.

———. 1992a. Extending Wittgenstein: The Pivotal Move from Epistemology to the Sociology of Science. In A. Pickering, ed., *Science as Practice and Culture* (Chicago and London: University of Chicago Press), pp. 215–65.

———. 1992b. From the "Will to Theory" to the Discursive Collage: Reply to Bloor. In A. Pickering, ed., *Science as Practice and Culture* (Chicago and London: University of Chicago Press), pp. 283–306.

———. 1992c. Springs of Action or Vocabularies of Motive. Paper presented at a workshop, "Vocation, Work, and Culture in Early Modern England," sponsored by the Achievement Project: Intellectual and Material Culture in Modern Europe, Oxford, 10–12 December.

———. 1993a. Representation Is Overrated: Some Critical Remarks about the Uses of the Concept of Representation in Science Studies. Presented at the conference "Located Knowledges: Intersections between Cultural, Gender, and Science Studies," University of California, Los Angeles, 8–10 April.

———. 1993b. *Scientific Practice and Ordinary Action: Ethnomethodology and Social Studies of Science.* Cambridge: Cambridge University Press.

Lynch, M., E. Livingston, and H. Garfinkel. 1983. Temporal Order in Laboratory Work. In K. Knorr-Cetina and M. Mulkay, eds., *Science Observed: Perspectives on the Social Study of Science* (Beverly Hills: Sage), pp. 205–38.

Lynch, M., and S. Woolgar, eds. 1990. *Representation in Scientific Practice.* Cambridge: MIT Press.

Lyotard, J.-F. 1984. *The Postmodern Condition: A Report on Knowledge.* Minneapolis: University of Minnesota Press.

———. 1988. *The Differend: Phrases in Dispute.* Minneapolis: University of Minnesota Press.

MacKenzie, D. 1981a. Interests, Positivism and History. *Social Studies of Science* 11:498–504.

———. 1981b. *Statistics in Britain, 1865–1930: The Social Construction of Scientific Knowledge.* Edinburgh: Edinburgh University Press.

MacKenzie, D., and J. Wajcman. 1985. The Social Shaping of Technology. In D. MacKenzie and J. Wajcman, eds., *The Social Shaping of Technology: How the Refrigerator Got Its Hum* (Milton Keynes, England: Open University Press), pp. 2–25.

McMullin, E., ed. 1988. *Construction and Constraint: The Shaping of Scientific Rationality.* Notre Dame, IN: University of Notre Dame Press.

Mahoney, M. S. Forthcoming. The Mathematical Realm of Nature. In D. Garber, ed., *The Cambridge History of Seventeenth-Century Philosophy.* Cambridge: Cambridge University Press.

Marinelli, M., and G. Morpurgo. 1982. Searches of Fractionally Charged Particles in Matter with the Magnetic Levitation Technique. *Physics Reports* 85:161–258.

———. 1984. The Electric Neutrality of Matter: A Summary. *Physics Letters* 137B:439–42.

Marinelli, M., G. Morpurgo, and G. L. Olcese. 1989. Diamagnetism of Powder and Bulk Superconducting YBCO Measured with a New Magnetic Levitometer. *Physica* C157:149–58.

Maturana, H. R., and F. J. Varela. 1992. *The Tree of Knowledge: The Biological Roots of Human Understanding.* Rev. ed. Boston: Shambala.

Megill, A. 1991. Four Senses of Objectivity. *Annals of Scholarship* 8:301–20.

———, ed. 1991–92. *Rethinking Objectivity.* Parts 1 and 2. Special issue of *Annals of Scholarship,* vol. 8 (3–4), vol. 9 (1–2).

Mellor, D. H. 1974. In Defence of Dispositions. *Philosophical Review* 83:157–81.

Messer-Davidow, E., D. Shumway, and D. Sylvan, eds. 1993. *Knowledges: Historical and Critical Studies in Disciplinarity.* Charlottesville: University Press of Virginia.

Miller, P. 1992. Accounting and Objectivity: The Invention of Calculating Selves and Calculable Spaces. In A. Megill, ed., *Rethinking Objectivity,* part 2, *Annals of Scholarship* 9 (1–2): 61–86.

Miller, P., and T. O'Leary. 1994. Accounting, "Economic Citizenship," and the Spatial Reordering of Manufacture. *Accounting, Organizations, and Society* 19:15–43.

Miller, P., and N. Rose. 1993. Production, Identity, and Democracy. London School of Economics. Manuscript.

Mishima, Y. 1973. *The Temple of Dawn.* New York: Knopf.

———. 1974. *The Decay of the Angel.* New York: Knopf.

Mol, A. 1991. Wombs, Pigmentation, and Pyramids: Should Antiracists and Feminists Try to Confine "Biology" to Its Proper Place? In J. L. Hermson and A. von Lenning, eds., *Sharing the Difference: Feminist Debates in Holland* (London: Routledge), pp. 149–63.

Mol, A., and J. Law. 1994. Regions, Networks, and Fluids: Anaemia and Social Topology. *Social Studies of Science* 24:641–71.

Morpurgo, G. 1965. Is a Non-relativistic Approximation Possible for the Internal Dynamics of "Elementary" Particles? *Physics* 2:95–105.

———. 1972. A Search for Quarks (a Modern Version of the Millikan Experiment): One Researcher's Personal Account. Genoa. Mimeo.

Morpurgo, G., G. Gallinaro, and G. Palmieri. 1970. The Magnetic Levitation Electrometer and Its Use in the Search for Fractionally Charged Particles. *Nuclear Instruments and Methods* 79:95–124.

Morse, P. M. 1977. *In at the Beginnings: A Physicist's Life.* Cambridge: MIT Press.

Mulkay, M. 1985. *The Word and the World: Explorations in the Form of Sociological Analysis.* London: George Allen and Unwin.

Mumford, L. 1934. *Technics and Civilization.* New York: Harcourt, Brace.

Murphy, N. 1993. Philosophical Fractals, or History as Metaphilosophy. *Studies in History and Philosophy of Science* 24:501–8.

Nagel, E. [1935] 1979. "Impossible Numbers": A Chapter in the History of Modern Logic. In *Teleology Revisited and Other Essays in the Philosophy and History of Science* (New York: Columbia University Press), pp. 166–94.

Newton-Smith, W. 1993. Science, Rationality, and Judgment. Presentation at the History and Philosophy of Science Departmental Colloquium, Cambridge, 4 March.

Nickles, T. 1992. Good Science as Bad History: From Order of Knowing to Order of Being. In E. McMullin, ed., *The Social Dimensions of Science* (Notre Dame, IN: University of Notre Dame Press), pp. 85–129.

Nicolson, M. 1991. The Social and the Cognitive: Resources for the Sociology of Scientific Knowledge. *Studies in the History and Philosophy of Science* 22:347–69.

Noble, D. F. 1979. *America by Design: Science, Technology, and the Rise of Corporate Capitalism.* Oxford: Oxford University Press.

———. 1986. *Forces of Production: A Social History of Industrial Automation.* Oxford: Oxford University Press.

O'Connell, J. 1993. Metrology: The Creation of Universality by the Circulation of Particulars. *Social Studies of Science* 23:129–73.

O'Neill, J. 1986. Formalism, Hamilton, and Complex Numbers. *Studies in History and Philosophy of Science* 17:351–72.

Peirce, C. S. [1891] 1923. The Architecture of Theories. In *Chance, Love, and Logic: Philosophical Essays,* edited by M. R. Cohen (New York: Harcourt, Brace), pp. 157–78.

Penley, C., and A. Ross, eds. 1991. *Technoculture.* Minneapolis: University of Minnesota Press.

Pestre, D., and J. Krige. 1992. Some Thoughts on the Early History of CERN. In P. Galison and B. Hevly, eds., *Big Science: The Growth of Large-Scale Research* (Stanford: Stanford University Press), pp. 78–99.

Piaget, J. 1985. *The Equilibration of Cognitive Structures: The Central Problem of Intellectual Development.* Chicago: University of Chicago Press.

Pickering, A. 1981a. Constraints on Controversy: The Case of the Magnetic Monopole. *Social Studies of Science* 11:63–93.

———. 1981b. The Hunting of the Quark. *Isis* 72:216–36.

———. 1981c. The Role of Interests in High-Energy Physics: The Choice between Charm and Colour. In K. D. Knorr, R. Krohn, and R. D. Whitley, eds., *Sociology of the Sciences,* vol. 4, *The Social Process of Scientific Investigation* (Dordrecht: Reidel), pp. 107–38.

———. 1984a. Against Putting the Phenomena First: The Discovery of the Weak Neutral Current. *Studies in History and Philosophy of Science* 15:85–117.
———. 1984b. *Constructing Quarks: A Sociological History of Particle Physics.* Chicago: University of Chicago Press.
———. 1986. Positivism/Holism/Constructivism. Paper presented at the weekly Colloquium of the Institute for Advanced Study, Princeton, 8 January 1987.
———. 1987. Forms of Life: Science, Contingency, and Harry Collins. *British Journal for History of Science* 20:213–21.
———. 1989a. Big Science as a Form of Life. In M. De Maria, M. Grilli, and F. Sebastiani, eds., *The Restructuring of the Physical Sciences in Europe and the United States, 1945–1960* (Singapore: World Scientific Publishing), pp. 42–54.
———. 1989b. Editing and Epistemology: Three Accounts of the Discovery of the Weak Neutral Current. In L. Hargens, R. A. Jones, and A. Pickering, eds., *Knowledge and Society: Studies in the Sociology of Science, Past and Present,* vol. 8 (Greenwich, CT: JAI Press), pp. 217–32.
———. 1989c. Living in the Material World: On Realism and Experimental Practice. In D. Gooding, T. J. Pinch, and S. Schaffer, eds., *The Uses of Experiment: Studies of Experimentation in the Natural Sciences* (Cambridge: Cambridge University Press), pp. 275–97.
———. 1990a. Knowledge, Practice, and Mere Construction. *Social Studies of Science* 20:682–729.
———. 1990b. Openness and Closure: On the Goals of Scientific Practice. In H. Le Grand, ed., *Experimental Inquiries: Historical, Philosophical, and Social Studies of Experimentation in Science* (Dordrecht: Kluwer), pp. 215–39.
———. 1991a. Objectivity and the Mangle of Practice. *Annals of Scholarship* 8:409–25.
———. 1991b. Philosophy Naturalised a Bit. *Social Studies of Science* 21 (3): 575–85.
———. 1991c. Reason Enough? More on Parity-Violation Experiments and Electroweak Gauge Theory. In A. Fine, M. Forbes, and L. Wessels, eds., *PSA 1990: Proceedings of the 1990 Biennial Meeting of the Philosophy of Science Association,* vol. 2, *Symposium and Invited Papers* (East Lansing, MI: Philosophy of Science Association), pp. 459–69.
———. 1992a. The Rad Lab and the World. Essay review of *Lawrence and His Laboratory: A History of the Lawrence Berkeley Laboratory,* vol. 1, by J. L. Heilbron and R. W. Seidel. *British Journal for History of Science* 25:247–51.
———. 1993a. Anti-Discipline or Narratives of Illusion. In E. Messer-Davidow, D. Shumway, and D. Sylvan, eds., *Knowledges: Historical and Critical Studies in Disciplinarity* (Charlottesville: University Press of Virginia), pp. 103–22.
———. 1993b. The Mangle of Practice: Agency and Emergence in the Sociology of Science. *American Journal of Sociology* 99:559–89.
———. 1994a. Beyond Constraint: The Temporality of Practice and the Historicity of Knowledge. In J. Buchwald, ed., *Scientific Practice: Theories and Stories of Physics* (Chicago: University of Chicago Press, 1995), pp. 42–55.
———. 1994b. Cyborg History and the WWII Regime. Paper presented at the

Davis Center Seminar, Princeton University, 25 March. To appear in *Perspectives on Science* 3(1).

———, ed. 1992b. *Science as Practice and Culture*. Chicago and London: University of Chicago Press.

Pickering, A., and A. Stephanides. 1992. Constructing Quaternions: On the Analysis of Conceptual Practice. In A. Pickering, ed., *Science as Practice and Culture* (Chicago and London: University of Chicago Press), pp. 139–67.

Pinch, T. J. 1977. What Does a Proof Do if It Does Not Prove? A Study of the Social Conditions and Metaphysical Divisions Leading to David Bohm and John von Neumann Failing to Communicate in Quantum Physics. In E. Mendelsohn, P. Weingart, and R. Whitley, eds., *The Social Production of Scientific Knowledge: Sociology of the Sciences* (Dordrecht: Reidel), 1:171–215.

———. 1985. Towards an Analysis of Scientific Observation: The Externality and Evidential Significance of Observational Reports in Physics. *Social Studies of Science* 15:3–36.

———. 1986. *Confronting Nature*. Dordrecht: Reidel.

Pinch, T. J., and W. Bijker. 1984. The Social Construction of Facts and Artefacts, or How the Sociology of Science and the Sociology of Technology Might Benefit Each Other. *Social Studies of Science* 14:399–441. Shortened and updated in W. Bijker, T. Hughes, and T. J. Pinch, eds., *The Social Construction of Technological Systems: New Directions in the Sociology and History of Technology* (Cambridge: MIT Press, 1987), pp. 17–50.

Poincaré, H. 1946. *The Foundations of Science*. Trans. G. B. Halsted. Lancaster, PA: Science Press.

Polanyi, M. 1958. *Personal Knowledge: Towards a Post-Critical Philosophy*. Chicago: University of Chicago Press.

Popper, K. [1935] 1959. *The Logic of Scientific Discovery*. New York: Basic Books.

———. [1957] 1986. *The Poverty of Historicism*. London: Routledge.

Porter, T. M. 1992a. Quantification and the Accounting Ideal in Science. *Social Studies of Science* 22:633–51.

———, ed. 1992b. Symposium on the Social History of Objectivity. *Social Studies of Science* 22:595–651.

Pycior, H. 1976. *The Role of Sir William Rowan Hamilton in the Development of Modern British Algebra*. Ph.D. dissertation, Cornell University.

Pynchon, T. 1975. *Gravity's Rainbow*. London: Picador.

Rasmussen, N. 1993. Facts, Artifacts, and Mesosomes: Practicing Epistemology with the Electron Microscope. *Studies in History and Philosophy of Science* 24:227–65.

Richards, R. J. 1987. *Darwin and the Emergence of Evolutionary Theories of Mind and Behavior*. Chicago and London: University of Chicago Press.

Richardson, G. R. 1991. *Feedback Thought in Social Science and Systems Theory*. Philadelphia: University of Pennsylvania Press.

Rorty, R. 1979. *Philosophy and the Mirror of Nature*. Princeton: Princeton University Press.

Rose, N. 1990. *Governing the Soul: The Shaping of the Private Self*. New York: Routledge.

Rose, N., and P. Miller. 1992. Political Power beyond the State: Problematics of Government. *British Journal of Sociology* 43:173–205.

Rouse, J. 1985. Heidegger's Later Philosophy of Science. *Southern Journal of Philosophy* 23:75–92.

———. 1986. Merleau-Ponty and the Existential Conception of Science. *Synthese* 66:249–72.

———. 1987a. Husserlian Phenomenology and Scientific Realism. *Philosophy of Science* 54:222–32.

———. 1987b. *Knowledge and Power: Toward a Political Philosophy of Science.* Ithaca: Cornell University Press.

———. 1992. What Are Cultural Studies of Scientific Knowledge? *Configurations* 1:1–22.

———. 1994. From Davidsonian Semantics to Cultural Studies of Science. Paper presented at a workshop, "The New 'Contextualism': Science as Discourse and Culture," University of Florida, 12–13 March.

———. Forthcoming. Engaging Science: Science Studies after Realism, Rationality, and Social Constructivism.

Rudwick, M. J. S. 1985. *The Great Devonian Controversy: The Shaping of Scientific Knowledge among Gentlemanly Specialists.* Chicago and London: University of Chicago Press.

Said, E. 1978. *Orientalism.* New York: Pantheon.

Schaffer, S. 1988. Astronomers Mark Time: Discipline and the Personal Equation. *Science in Context* 2:115–45.

———. 1991. The Eighteenth Brumaire of Bruno Latour. *Studies in History and Philosophy of Science* 22:174–92.

———. 1992a. Late Victorian Metrology and Its Instrumentation: A Manufactory of Ohms. In R. Bud and S. E. Cozzens, eds., *Invisible Connections: Instruments, Institutions, and Science* (Washington, DC: SPIE Optical Engineering Press), pp. 23–56.

———. 1992b. Self Evidence. *Critical Inquiry* 18:327–62.

———. 1993a. Babbage's Intelligence: Calculating Engines and the Factory System. University of Cambridge. Manuscript.

———. 1993b. Empires of Physics. Paper presented at a conference, "Empires and Knowledge," University of Cambridge, 13 March.

Schweber, S. S. 1992a. Big Science in Context: Cornell and MIT. In P. Galison and B. Hevly, eds., *Big Science: The Growth of Large-Scale Research* (Stanford: Stanford University Press), pp. 149–83.

———. 1992b. A Historical Perspective on the Rise of the Standard Model: Context, Traditions, Men. Paper presented at the conference on the history of the standard model in particle physics, Stanford, CA, June.

Secord, J. A., ed. 1993. *The Big Picture.* Special issue of *British Journal for History of Science* 36 (4): 385–483.

Serres, M. 1982. *Hermes: Literature, Science, Philosophy.* Baltimore and London: Johns Hopkins University Press.

Sewell, W. H. 1992. A Theory of Structure: Duality, Agency, and Transformation. *American Journal of Sociology* 98:1–29.

Shapere, D. 1982. The Concept of Observation in Science and Philosophy. *Philosophy of Science* 49:485–525.

Shapin, S. 1979. The Politics of Observation: Cerebral Anatomy and Social Interests in the Edinburgh Phrenology Disputes. In R. Wallis, ed., *On the Margins of Science: The Social Construction of Rejected Knowledge,* Sociological Review Monograph 27 (Keele, England: University of Keele), pp. 139–78.

———. 1982. History of Science and Its Sociological Reconstructions. *History of Science* 20:157–211.

———. 1988a. Following Scientists Around. *Social Studies of Science* 18:533–50.

———. 1988b. The House of Experiment in Seventeenth-Century England. *Isis* 79:373–404.

———. 1991. Discipline and Bounding: The History and Sociology of Science as Seen through the Externalism-Internalism Debate. In *Critical Problems and Research Frontiers in History of Science and History of Technology,* papers for a conference held in Madison, WI, 30 October–3 November, pp. 203–27.

Shapin, S., and S. Schaffer. 1985. *Leviathan and the Air Pump: Hobbes, Boyle, and the Experimental Life.* Princeton: Princeton University Press.

Sibum, H. O. 1992. Reworking the Mechanical Value of Heat: Instruments of Precision and Gestures of Accuracy in Early Victorian England. Paper presented at a workshop, "Replications of Historical Experiments in Physics," Carl von Ossietzky University, Oldenburg, Germany, 24–29 August. To appear, with revision, in *Studies in History and Philosophy of Science.*

Singleton, V., and M. Michael. 1993. Actor-Networks and Ambivalence: General Practitioners in the UK Cervical Screening Programme. *Social Studies of Science* 23:227–64.

Slätis, H. 1959. On Bubble Chambers. *Nuclear Instruments and Methods* 5:1–25.

Smith, B. H. 1988. *Contingencies of Value: Alternative Perspectives for Critical Theory.* Cambridge: Harvard University Press.

———. 1992. The Unquiet Judge: Activism without Objectivism in Law and Politics. *Annals of Scholarship* 9:111–33.

———. 1993a. Doing without Meaning. Duke University. Manuscript.

———. 1993b. Unloading the Self-Refutation Charge. *Common Knowledge* 2:81–95.

Smith, C., and M. N. Wise. 1989. *Energy and Empire: A Biographical Study of Lord Kelvin.* Cambridge: Cambridge University Press.

Star, S. L. 1991a. Power, Technology, and the Phenomenology of Conventions: On Being Allergic to Onions. In J. Law, ed., *A Sociology of Monsters? Essays on Power, Technology, and Domination* (London: Routledge), pp. 27–57.

———. 1991b. The Sociology of the Invisible: The Primacy of Work in the Writings of Anselm Strauss. In D. R. Maines, ed., *Social Organization and Social Process: Essays in Honor of Anselm Strauss* (Hawthorne, NY: Aldine de Gruyter), pp. 265–83.

———. 1992. The Trojan Door: Organizations, Work, and the "Open Black Box." *Systems/Practice* 5:395–410.

Star, S. L., and J. R. Griesemer. 1989. Institutional Ecology, "Translations," and Boundary Objects: Amateurs and Professionals in Berkeley's Museum of Vertebrate Zoology. *Social Studies of Science* 19:387–420.

Stover, R. W., T. Moran, and J. Trischka. 1967. Search for an Electron-Proton Charge Inequality by Charge Measurements on an Isolated Macroscopic Body. *Physical Review* 164:1599–1609.

Strathern, M. 1991. *Partial Connections*. Savage, MD: Rowman and Littlefield.

Suchman, L. 1987. *Plans and Situated Actions: The Problem of Human-Machine Communication*. Cambridge: Cambridge University Press.

Suppe, F., ed., 1977. *The Structure of Scientific Theories*. 2d ed. Urbana: University of Illinois Press. 1st ed. published 1974.

Swatez, G. M. 1966. Social Organization of a University Laboratory. Internal Working Paper 44, Space Sciences Laboratory, Social Sciences Project, University of California, Berkeley.

———. 1970. The Social Organization of a University Laboratory. *Minerva* 8 (1): 37–58.

Taussig, M. 1987. *Shamanism, Colonialism, and the Wild Man: A Study in Terror and Healing*. Chicago: University of Chicago Press.

Thackray, A., ed. 1992. *Science after '40*. Special issue of *Osiris* 7.

Thompson, E. P. 1963. *The Making of the English Working Class*. New York: Penguin.

Tiles, M. 1984. *Bachelard: Science and Objectivity*. Cambridge: Cambridge University Press.

Traweek, S. 1988. *Beamtimes and Lifetimes: The World of High Energy Physicists*. Cambridge and London: Harvard University Press.

———. 1992. Border Crossings: Narrative Strategies in Science Studies and among Physicists in Tsukuba Science City, Japan. In A. Pickering, ed., *Science as Practice and Culture* (Chicago and London: University of Chicago Press), pp. 429–65.

Trenn, T. J. 1986. The Geiger-Müller Counter of 1928. *Annals of Science* 43:111–35.

Trower, W. P. 1989. Luis Walter Alvarez (1911–1988). In M. De Maria, M. Grilli, and F. Sebastiani, eds., *The Restructuring of the Physical Sciences in Europe and the United States, 1945–1960* (Singapore: World Scientific Publishing), pp. 105–15.

———, ed. 1987. *Discovering Alvarez: Selected Works of Luis W. Alvarez with Commentary by His Students and Colleagues*. Chicago and London: University of Chicago Press.

Turner, S. 1991. Two Theorists of Action: Ihering and Weber. *Analyse und Kritik* 13:46–60.

———. 1994. *The Social Theory of Practices: Tradition, Tacit Knowledge, and Presuppositions*. Chicago: University of Chicago Press.

van den Belt, H., and A. Rip. 1987. The Nelson-Winter-Dosi Model and Synthetic Dye Chemistry. In W. Bijker, T. Hughes, and T. J. Pinch, eds., *The Social Construction of Technological Systems: New Directions in the Sociology and History of Technology* (Cambridge: MIT Press), pp. 135–58.

van der Waerden, B. L. 1976. Hamilton's Discovery of Quaternions. *Mathematics Magazine* 49:227–34.

Voss, M. 1992. Niccolo Tartaglia: A Maestro d'Abbaco between the Cannon and the Book. Paper presented at the History of Science Society meeting, Washington, DC, December.

Webster, C. 1982. Paracelsus and Demons: Science as a Synthesis of Popular Belief. In L. S. Olschki, ed., *Scienze, Credenze Occulte, Livelli di Cultura* (Florence: Istituto Nazionale di Studi sul Rinascimento), pp. 3–20.

Weil, S. 1978. *Lectures on Philosophy.* Cambridge: Cambridge University Press.

Weiss, L. 1990. The Construction of CERN's First Hydrogen Bubble Chambers. In A. Hermann, J. Krige, U. Mersits, and D. Pestre, *History of CERN,* vol. 2, *Building and Running the Laboratory, 1954–1965* (Amsterdam: North-Holland), pp. 269–338.

Whittaker, E. T. 1945. The Sequence of Ideas in the Discovery of Quaternions. In Royal Irish Academy, *Proceedings* 50, sec. A, no. 6, pp. 93–98.

Wilson, A., ed. 1993. *Rethinking Social History: English Society 1570–1920 and Its Interpretation.* Manchester: Manchester University Press.

Wilson, B. R., ed. 1970. *Rationality.* Oxford: Blackwell.

Wise, M. N. 1988. Mediating Machines. *Science in Context* 2:77–113.

———. 1993. Mediations: Enlightenment Balancing Acts, or the Technologies of Rationalism. In P. Horwich, ed., *World Changes: Thomas Kuhn and the Nature of Science* (Cambridge: MIT Press), pp. 207–56.

———, ed. 1995. *The Values of Precision.* Princeton: Princeton University Press.

Wise, N., and C. Smith. 1989–90. Work and Waste. *History of Science* 27:263–301, 391–449, 28:221–61.

Wittgenstein, L. 1953. *Philosophical Investigations.* New York: Macmillan.

———. 1976. *Wittgenstein's Lectures on the Foundations of Mathematics, Cambridge, 1939.* Edited by C. Diamond. Hassocks, Sussex: Harvester Press.

Woolgar, S. 1988a. Reflexivity Is the Ethnographer of the Text. In *Science: The Very Idea* (London: Tavistock), pp. 14–34.

———. 1988b. *Science: The Very Idea.* London: Tavistock.

———. 1992. Some Remarks about Positionism: A Reply to Collins and Yearley. In A. Pickering, ed., *Science as Practice and Culture* (Chicago and London: University of Chicago Press), pp. 327–42.

———, ed. 1988c. *Knowledge and Reflexivity: New Frontiers in the Sociology of Knowledge.* Beverly Hills: Sage.

Zuboff, S. 1987. *In the Age of the Smart Machine.* New York: Basic.

Zuckerman, H. 1988. The Sociology of Science. In N. J. Smelser, ed., *Handbook of Sociology* (Beverly Hills: Sage), pp. 511–74.

Index